I0463818

The Papers of Independent Authors

Volume 22

Russia-Israel
2013

It is published **26.02.2013** (version 3)
It is published **25.12.2012** (version 2)
It is published **16.12.2012** (version 1)
It is sent in a press on Februar, 4th 2013
It is printed in the USA, Lulu Inc., catalogue **13499083**
ISBN 978-1-300-53127-2
EAN-13 9772225671006
ISSN 2225-6717
Site with information for the author - http://izdatelstwo.com
The contact information - publisher-dna@hotmail.com
Fax: ++972-8-8691348
Address: POB 15302, Bene-Ayish, Israel, 60860

ISBN 978-1-300-53127-2

9 781300 531272

90000

Russia-Israel
2013

Truth - the daughter of time, instead of authority

|| Frensis Bacon

Everyone has the right to freedom of opinion and expression; this right includes freedom to hold opinions without interference and to seek, receive and impart information and ideas through any media and regardless of frontiers.

United Nations OrganizationUniversal. Declaration of Human Rights. Article 19.

From the Publisher

"The Papers of Independent Authors" - versatile scientific and technical printed magazine. The magazine accepts articles to the publication from all countries. The following rules are complied by this:

- Articles are not reviewed and the publishing house is not responsible for a content and style of publications;
- The magazine is registered in the international qualifier of books ISBN, transferred and registered in national libraries of Russia, USA, Israel;
- The priority and copyrights of the author of paper are provided with registration of magazine in ISBN;
- The commercial rights of the author of paper are kept for the author;
- The magazine is published in Israel and printed in the USA;
- The printed magazine is on sale under the cost price (for publishing house), and in an electronic kind provides free of charge;
- The author pays for the publication.

This magazine - for those authors who are assured of himselves and do not require approval of the reviewer. Us often reproach that papers are not reviewed. But the institute of reviewing is not the ideal filter – it passes unsuccessful papers and detains original works. Without analyzing the numerous reasons of it, we shall notice only, that if the reader outstanding ideas can remain unknown persons can filter bad papers. Therefore we wish to give to scientists and to engineers the right (similarly to writers and artists) to be published without reviewing and to not to spend years for the statement of the ideas.

Solomon I. Khmelnik.

Contents

Series: **ELECTRICAL ENGINEERING**

Khmelnik S.I.

Calculation of electric direct current circuits with diodes

Annotation

In [1] has been described (in particular) method for calculation of electric direct current circuits with diodes. Here are given programs in the MATLAB with open source to solve this problem.

Contents

1. Introduction

In [1] describes (in particular) the method of calculation of electrical direct current circuits with diodes. There are open source MATLAB-programs for such calculations. Data for the calculation prepared in tabular form. Dimension of the task is limited only by computer resources.

2. Description of electric circuit

Initially, the electric circuit nodes are numbered in random order. Branches electric circuits are determined by the numbers **N1** initial and **N2** end node. A branch may contain a resistance **R**, a constant voltage source **U** and a diode **D**. The choice of which of the two nodes to designate initial, has value only if the branch contains a diode. The initial node is assigned one that is adjacent to the positive pole of the diode.

Current (defined as a result of the calculation) has a positive direction from the initial to the final node.

Description of the electric circuit is an array of **B**. Each row of the array describes one of the branches and has the following form:

$$B(k,:)=[N1,N2,R,U,D]$$

In this case, **D=1**, if the diode is in the branch, and **D=0** otherwise.

In addition, the electric circuit may include current sources, are included among the total points and some node. The positive direction of current from the source is sending to the node.

Description of current sources is an array following form:

$$C=[C1,C2,...,CN,...],$$

where each node **N** we associate the number of **CN** - the current value of the current source, or zero if in this node is not a current source.

3. On the nodal current sources

In a real electric circuit current source is included in some branch. To bring this electric circuit to the "canonical" form (described above), do the following:

   o converts the source of the current in a separate branch (not containing other elements)

   o convert a branch in the two current sources, as in the electric circuit shown in Fig. 1 - see the transformation **a--->b**.

Obviously, the canonical electric circuit must satisfy the condition

$$sum(C)=0.$$

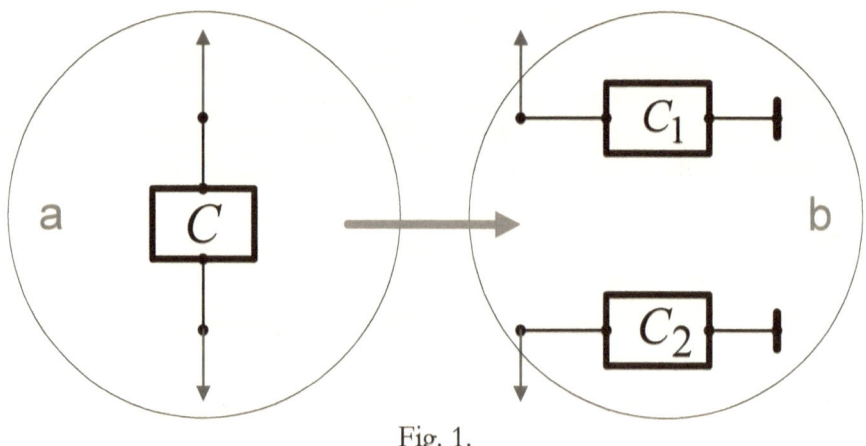

Fig. 1.

4. Calculation electric circuit

The calculation electric circuit is performed iteratively and the result of calculation, as a rule, is approximate. In this case, due to appear residuals - the deviation from zero in the equations of Kirchhoff's laws. They correspond to the relative error violations of these laws. The relative error violations the first law of Kirchhoff defined as the ratio of the mean square residuals of the first Kirchhoff's law to the mean square of the currents in the branches, and the relative error violations of the second law of Kirchhoff defined as the ratio of the mean square residuals for the second Kirchhoff's law to the mean square of the voltage on the branches (created voltage sources and current sources).

The value of the permissible relative error in violation of the second Kirchhoff's law is given by the user.

The number of iterations (i.e. duration of the calculation) and the error performance of the first law Kirchhoff adjust the amount of so-called "methodical" resistance. It makes sense the resistance included between each node and a common point. This resistance must be much greater than all the resistance branches (not counting reverse resistance diodes). The greater this resistance, the higher the accuracy of compliance of the first law of Kirchhoff, but the longer the duration of the calculations.

5. Program Description

M-function for the calculation is as follows:

```
function [i,f,er1,er2,k,p,E,N,y,m]=...
    rucd(B,C,r,erd,dmin,dmax,n)
```

Input arguments here are

B   —array   of   branches   (described   above),

C   –   array   of   current   sources   (described   above),

r   –   "methodical"   resistance,

erd – permissible given the magnitude of relative error violations of the   second   law   of   Kirchhoff,

dmin   –   resistance   of   the   diode   for   direct   current,

dmax   -   resistance   of   the   diode   for   reverse   current,

n - number of nodes.

Output values here are

i   —an array of current of branches

f —the array of nodal potentials

---

er1 – relative error violations the first law of Kirchhoff,

er2 – relative error violations of the second law of Kirchhoff,

k – the number of iterations

p –the array of residuals in the branches of the second law of Kirchhoff,

E – the array of potential difference between the nodes of branches,

N – the array - incidence matrix,

y – the array of residuals in the nodes of the first law of Kirchhoff,

m – a flag of the result, where

- **m=0**, if the calculation is made;
- **m=1**, if the calculation is not carried out due to violations of conditions of **sum(C)=0**; then get a message **msg=sum(C);**
- **m=2**, if in table nodes met the node number, exceeding a specified number of nodes; then get a message is **msg='greatest number'.**

6. Example of a "canonical" form

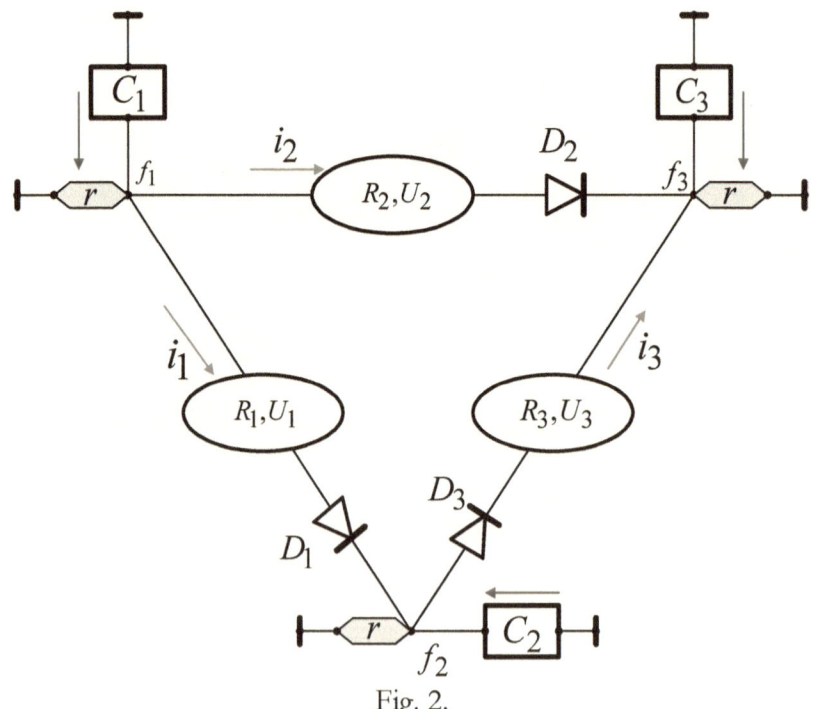

Fig. 2.

Fig. 2 shows a simple electric circuit with diodes. Here **r** – methodical resistance, which is absent in the real electric circuit. For this electric circuit these arrays are as follows:

```
B=[1,2,R1,U1,D1;...
   1,3,R2,U2,D2;...
   2,3,R3,U3,D3];
C=[C1,C2,C3]';
```

7. Test

Test M-function

```
function test()
```

calculates the electric circuit shown in Fig. 3.

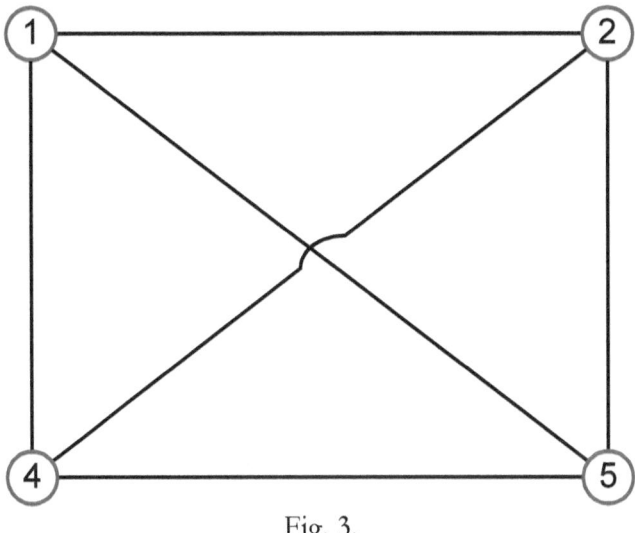

Fig. 3.

The presence of certain elements of the branches and sources of current determined in the arrays **B,C**.

8. References

1. Khmelnik S.I. Electrical DC electric circuit for modeling and control. Algorithms and Hardware. Published by "MiC" - Mathematics in Computer Comp., second edition, printed in USA, Lulu Inc., ID 293359, 2006, 177 c. (in Russian)

9. Programs

List of programs

1. function **test()**
2. function **[i,f,er1,er2,k,p,E,N,y,m]**
   =rucd(b,C,ro,erd,dmin,dmax,n)
3. function **[N,nuz] = makingN(branch)**
4. function **[R,U,D] = makingRU(branch)**
5. function **[f,i,k,er,erk,p,Un]=**
   nocondel(R,ro,N,U,C,erd,D,dmin,dmax)
6. function **[i,k,er,p]=**
   min2_4_3(Rn,Un,erd,D,dmin,dmax)
7. function **[mui1,mui2]=Rdiod(i,D,dmin,dmax)**

```
function test()
% B: begN, endN, R, U, D
B(1,:) =[ 2,1, 10,    0,        0 ];
B(2,:) =[ 2,3, 20,    0 ,       0 ];
B(3,:) =[ 3,4, 30,   -230,      0 ];
B(4,:) =[ 4,1, 40,    22 ,      0 ];
B(5,:) =[ 1,3, 50,    0,        1 ];
B(6,:) =[ 2,4, 60,    0,        1 ];
C=1*[0,3,-2,-1]';
dmin=0.001;
dmax=10000;
r=100000;
erd=0.0001;
n=4;
[i,f,er1,er2,k,p,E,N,y,m]=rucd(B,C,r,erd,dmin,d
max,n);
er1
er2
k
m


function [i,f,er1,er2,k,p,E,N,y,m]
=rucd(b,C,ro,erd,dmin,dmax,n,kmax)
% main function
if sum(C)==0
    m=0;
else
```

```
    m=1;
    msg=sum(C)
    i=0;f=0;er1=0;er2=0;k=0;p=0;E=0;N=0;y=0;
    return
end
[N,nuz] = makingN(b);
if nuz>n
    m=2;
    msg=nuz
    i=0;f=0;er1=0;er2=0;k=0;p=0;E=0;N=0;y=0;
    return
end
[R,U,D] = makingRU(b);
[f,i,k,er2,er1,p,E]=nocondel(R,ro,N,U,C,erd,D,d
min,dmax,kmax);
y=f/ro;


function [N,nuz] = makingN(branch)
% the creation of the incidence matrix
% b = begN, endN
    raz=size(branch);
    nb=raz(1);
    uz = 0;
    k = 1;
    while k <= nb
        a = branch(k,1);
        if a > uz
            uz = a;
        end
        b = branch(k,2);
        if b > uz
            uz = b;
        end
        if a > 0
            N(a , k) = -1;
        end
        if b > 0
            N(b , k) = 1;
        end
        k = k + 1;
    end
```

```
      nuz=uz;
      n=size(N);
      nuz2=n(1);
      if nuz==nuz2
      else
           nuz
           nuz2
           msg='Numbers of Nodes?'
      end
```

function [R,U,D] = makingRU(branch)

```
%   R,U,D from branches
%   b = uzbegN,uzendN,R,U,D
      raz=size(branch);
      nb=raz(1);
      uz = 0;
      k = 1;
      while k <= nb
           R(k,k) = branch(k,3);
           U(k)   = branch(k,4);
           D(k)   = branch(k,5);
           k=k+1;
      end
      U=U';
```

function [f,i,k,er,erk,p,Un]=... nocondel(R,ro,N,U,C,erd,D,dmin,dmax,kmax)

```
% basic calculations
Rn=R+ro*N'*N;
Un=U-ro*N'*C;
[i,k,er,p]=min2_4_3(Rn,Un,erd,D,dmin,dmax,kmax)
;
f=ro*(N*i+C);
raz=size(N);
nf=sum(f.^2)/raz(1);
ni=sum(i.^2)/raz(2);
erk=sqrt(nf/(ro*ro*ni));
```

function [i,k,er,p]=... min2_4_3(Rn,Un,erd,D,dmin,dmax,kmax)

```
% min. function (1.1.24)
```

```
i=0*Un;
nUn=sum((Un).^2);
er=999;
k=0;
while er>erd && k<kmax
    k=k+1;
    [mui1,mui2]=Rdiod(i,D,dmin,dmax);
    p=mui1'+Rn*i-Un;
    np=sum((p).^2);
    er=sqrt(np/nUn);
    if er==0
        break;
    end
    m2=diag(mui2);
    a=p'*p/(p'*(m2+Rn)*p);
    ap=a*p;
    i=i-ap;
end

function [mui1,mui2]=Rdiod(i,D,dmin,dmax)
% resistance of diodes
n=size(i);
n=n(1);
k=1;
while k<=n
    if D(k)==1
        if i(k) >0
            mui1(k)=dmin*i(k);
            mui2(k)=dmin;
        else % i(k) =<0
            mui1(k)=dmax*i(k);
            mui2(k)=dmax;
        end
    else
        mui1(k)=0;
        mui2(k)=0;
    end
    k=k+1;
end
```

Series: **HYDRODYNAMICS**

Khmelnik S.I.

The Emergence Mechanism and Calculation Method of Turbulent Flows

Abstract

An explanation is offered of the mechanism of turbulent flows emergence, based of the Maxwell-like gravitation equations, updated after some known experiments. It is shown that the moving molecules of flowing liquid interact like electrical charges.

The forces of such interaction can be calculated and included to the Navier-Stokes equations as mass forces. Navier-Stocks equations complemented by these forces become equations of hydrodynamics for turbulent flow. For the calculations of turbulent flows the known methods of Navier-Stokes equations solution may be used.

Contents

1. Introduction

In [1] an analogy of electromagnetism and gravito-electromagnetism has been explored. From the point of this analogy the new experiments of Samokhvalov [2] have been analyzed. Based on this analysis it was shown there, that Maxwell-like equations of gravito-electromagnetism should be supplemented by a certain empirical coefficient of gravitational permeability of the medium. For vacuum this coefficient is about

$\xi \approx 10^{12}$, and it decreases rapidly with the pressure increase. This explains the absence of visual effects of gravito-magnetic interaction of moving masses in the air. However in vacuum theses interactions are clearly visible in the above mentioned experiments.

In the liquid flow the moving molecules are separated by vacuum. So their gravito-magnetic interaction forces can be substantial and influence the nature of the flow.

We know that with increasing speed of laminar liquid or gas turbulence may occur spontaneously (without the influence of external forces) [3]. The mechanism for turning from laminar to turbulent flow has not been found. Evidently, a source of forces perpendicular to the flow speed must be found.

Further it is shown that the gravito-magnetic interaction of the moving liquid masses can be the cause of the turbulence emergence.

2. The Interaction of Moving Electrical Charges

Let us consider two charges q_1 and q_2, moving with speeds v_1 and v_2 accordingly. It is known [4], that the induction of the field created by the charge q_1 in the point in which the charge q_2 is located, is equal (here and further we are using the CGS System)

$$\overline{B_1} = q_1 \left(\overline{v_1} \times \overline{r} \right) / cr^3 . \tag{1}$$

Here the vector $\overline{r}$ is directed from the point where the moving charge q_1 is located. The Lorentz force acting on the charge q_2, is

$$\overline{F_{12}} = q_2 \left(\overline{v_2} \times \overline{B_1} \right) / c . \tag{2}$$

Similarly,

$$\overline{B_2} = q_2 \left(\overline{v_2} \times \overline{r} \right) / cr^3 , \tag{3}$$

$$\overline{F_{21}} = q_1 \left(\overline{v_1} \times \overline{B_2} \right) / c . \tag{4}$$

In the general case $\overline{F_{12}} \neq \overline{F_{21}}$, i.e. the third Newton law does not work – there are unbalanced forces acting on the charges q_1 and q_2 and bending the trajectories of these charges movement.

Let us consider the correlation between the Lorentz force and the force of charges attraction. In the simplest case the Lorentz force found from (1, 2) is

$$F = \frac{q_1 q_2 v_1 v_2}{r^2 c^2}. \tag{5}$$

The force of attraction of the two charges is

$$P = \frac{q_1 q_2}{r^2}. \tag{6}$$

Consequently,

$$\phi_e = \frac{F}{P} = \frac{v_1 v_2}{c^2}. \tag{7}$$

We shall call this magnitude the efficiency of Lorentz forces.

3. Gravito-Magnetic Interaction of Moving Masses

In analogy with the electrical charges interaction, two masses m_1 and m_2, moving with speeds v_1 and v_2 accordingly are also interacting. In [1] it is shown that in this case there emerge gravito-magnetic inductions of the form:

$$\overline{B_{g1}} = G m_1 \left(\overline{v_1} \times \overline{r} \right) / c r^3 , \tag{1}$$

$$\overline{B_{g2}} = G m_2 \left(\overline{v_2} \times \overline{r} \right) / c r^3 , \tag{2}$$

where

c — the speed of light in vacuum, $c \approx 3 \cdot 10^{10}$ [cm/sec];

G - gravitational constant, $G \approx 7 \cdot 10^{-8}$ [dynes·cm²·g⁻²].

The masses are also affected by the Lorentz gravito-magnetic Lorentz forces of the following form [1]:

$$\overline{F_{12}} = \varsigma \xi\, m_2 \left(\overline{v_2} \times \overline{B_{g1}} \right) c , \tag{3}$$

$$\overline{F_{21}} = \varsigma \xi m_1 \left(\overline{v_1} \times \overline{B_{g2}} \right) c , \tag{4}$$

where

$\varsigma = 2$, which follows from GRT,

$\xi \approx 10^{12}$ - the gravitational permeability coefficient for vacuum [1].

For parallel speeds and equal mass forces $\overline{F_{12}} = -\overline{F_{21}}$ the laminar flow keeps its character. However, in the general case, when $v_1 \neq v_2$, the forces $\overline{F_{12}} \neq \overline{F_{21}}$ are generated, i.e. the unbalanced force

$\Delta F = \overline{F}_{12} + \overline{F}_{21}$, acting on the masses m_1 and m_2 and bending the trajectories of these masses movement (let us note that here the Newton's third law is not observed [4]). From the above formulas follows that the unbalanced force is directed at an angle to the flow speed, which violates the laminarity.

Let us find the correlation between the Lorentz gravito-magnetic force and the masses attraction force. Similarly to the previous, in the simplest case Lorentz gravito-magnetic force may be found from (1, 3):

$$F = \varsigma\xi \frac{Gm_1m_2v_1v_2}{r^2c^2}.$$ (5)

The attraction force of two masses is

$$P = \frac{Gm_1m_2}{r^2}.$$ (6)

Thus,

$$\phi_g = \frac{F}{P} = \varsigma\xi \cdot \frac{v_1v_2}{c^2}.$$ (7)

We shall call this magnitude – the Lorentz gravito-magnetic force efficiency Comparing (2.7) and (3.7) we find that

$$\phi_g = \phi_e\varsigma\xi.$$ (8)

Consequently, the efficiency of Lorentz gravito-magnetic forces is much higher than Lorentz electromagnetic forces efficiency for comparable speeds.

4. Gravito-Magnetic Interaction as the Cause of Turbulence

For the appearance of unbalanced forces the following conditions must be satisfied:
1. the speeds must be of certain magnitude (which make these forces substantial);
2. there must be a cause for local change of the speeds, for instance
   o an appearance of a barrier,
   o a change of pressure when a stream flows from the water.

There may be a number of reasons increasing the unbalanced forces:
- Temperature rise causing the speeds v_1 and v_2 to cease being parallel due to heat fluctuations,

- Viscosity reduction, i.e. the reduction of intermolecular attraction forces which counteract the unbalanced forces that take   the molecules apart.

We may specify also a number of external factors causing the appearance of unbalanced forces due to external violation of speeds v_1 and v_2 parallelism, for instance:

- abrupt changes in temperature, pressure;
- the injection of extra liquid or other agent.

A local change of equal speeds of a pair of linked molecules, caused, for instance, by asymmetric blow, is inevitably spread to the whole flow area.

As the Lorentz forces do not perform any work, the energy for turbulent motion must be supplied from the energy of laminar flow, which means that the energy of input flow must exceed a certain magnitude for the turbulence appearance.

The Navier-Stokes equations permit to determine of a speed of flow that meets a barrier or leaves a barrier. Knowing these speeds, we may determine unbalanced forces from the said equations. Then these forces as the functions of speeds, may be included into the Navier-Stokes equations as the mass forces.

5. Quantitative Estimates

In the general case, from (3.2, 3.4) we can find

$$\overline{F_{21}} = \frac{\varsigma\xi G m_1 m_2}{c^2 r^3}\left(\overline{v_1} \times \left(\overline{v_2} \times \overline{r}\right)\right). \tag{1}$$

Let us consider the vectors' orts, denoting them by a stroke. Then from (1) we get:

$$\overline{F_{21}} = \sigma \overline{f_{21}}, \tag{2}$$

where

$$\overline{f_{21}} = \left(\overline{v_1'} \times \left(\overline{v_2'} \times \overline{r'}\right)\right). \tag{3}$$

$$\sigma = \frac{\varsigma\xi G \cdot m_1 m_2 v_1 v_2}{c^2 r^2}, \tag{4}$$

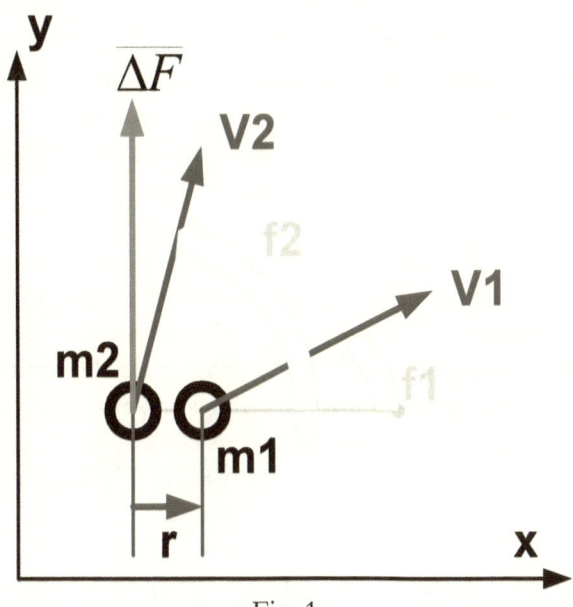

Fig. 1.

In the same way,

$$\overline{F_{12}} = \sigma \overline{f_{12}},\tag{5}$$

where

$$\overline{f_{12}} = \left(\overline{v_2'} \times \left(\overline{v_1'} \times \overline{r'}\right)\right),\tag{6}$$

and

$$\overline{\Delta F} = \sigma \overline{\Delta f},\tag{7}$$

where

$$\overline{\Delta F} = \overline{F_{21}} + \overline{F_{12}},\tag{8}$$

$$\overline{\Delta f} = \overline{f_{21}} + \overline{f_{12}}.\tag{9}$$

Let us consider two adjacent molecules of the liquid. The distance between the molecules of liquid stays invariable. Due to the smallness of distance r between them, we may assume that the vectors of speeds $\overline{v_1'}$, $\overline{v_2'}$ of these molecules are applied to one point and lie in the same plane xoy. Then vector (9) also lies in this plane. Fig. 1 shows the position of vectors $\overline{v_1'}$, $\overline{v_2'}$, $\overline{r'}$.

In the Supplement (see (6)) is shown that the magnitude of vector (9) is given by formula

$$\Delta f = r \sin(\varphi_2 - \varphi_1).\tag{8}$$

Taking into account (9, 10) we shall get:

$$\Delta F = \sigma \sin(\varphi_2 - \varphi_1).$$ (9)

This force appears when the adjacent molecules hit the barrier under different angles. We may assume that the summary force is applied to one of the molecules. Thus it creates a torque of dipole consisting of two molecules,

$$M = r \cdot \Delta F.$$ (10)

Each pair of adjacent molecules generates a dipole with a torque (10). The torques increase the local speeds of the liquid molecules, which, in its turn, increase the torques of the said dipoles. Because of this, the turbulence once began continues to grow, spreading in the liquid volume. Formula (9) determines the forces of gravito-magnetic interaction of the liquid molecules as a function of speeds of these contacting molecules.

These forces can be included to the Navier-Stokes equations as mass forces – see further.

6. Example: Turbulent Water Flow in a Pipe

Now we shall consider the case of interaction between liquid streams, assuming that the interaction is between groups of molecules forming an element of a stream. We shall take a specific case when the speed vectors of the streams are the same $|v_1| = |v_2| = v$ and also the masses of these groups are $m_1 = m_2 = m$. From (4) we can find

$$\sigma = \varsigma \xi G \left(\frac{mv}{cr} \right)^2.$$ (11)

where r is the distance between the streams. Let us denote as d a typical size of the group (the stream diameter) and rewrite (11) as

$$\sigma = \varsigma \xi G \left(\frac{\rho \cdot d^3 v}{cr} \right)^2.$$ (11a)

where ρ is liquid density, and the mass of the group is

$$m = \rho \cdot d^3.$$ (11в)

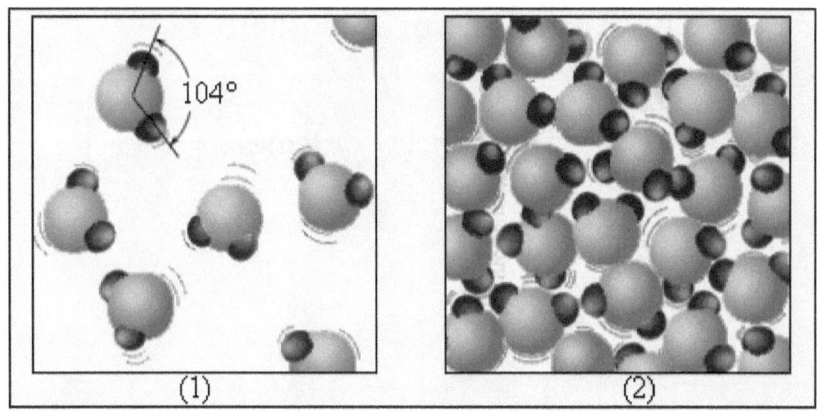

Fig. 2 (from Vikipedia). Water vapor (1) and water (2). Molecules of water are enlarged by $5 \cdot 10^7$ times.

The further example is related to water. As in liquids the molecules are located at a distance commensurable with the size of molecule itself (see Fig. 2), we shall take the distance between molecules equal to the molecule diameter, w2which for water is $r \approx 3 \cdot 10^{-12} [cm]$. The water density is $\rho = 1 [g/cm^3]$ Let us find also the speed of water flow where the turbulence occurs. It is known [3], that the condition of turbulence occurrence is given by Reinolds criterion, which for a round pipe is

$$\text{Re} = Dv / \eta, \tag{12}$$

where D is the pipe's diameter, η is the kinematic viscosity coefficient – for water it is $\eta \approx 0.01 [cm^2/\text{sec}]$ [5]. Let $D = 2.5 [cm]$. The turbulence occurs when the Reinolds number is $\text{Re} > 2300$. Now from (12) let us find the speed of turbulent flow: $v = 10$ [cm/sec] Let the diameter of interacting streams $d \approx 0.1 [cm]$. It was mentioned above that $\varsigma = 2$, $\xi \approx 10^{12}$, $G \approx 7 \cdot 10^{-8}$. Then from (11a) we find

$$\sigma = 2 \cdot 10^{12} \cdot 7 \cdot 10^{-8} \left(1 \cdot 0.1^3 \cdot 10 / \left(3 \cdot 10^{10} \cdot 3 \cdot 10^{-12} \right) \right) \approx 2000 \text{ [dynes] (13)}$$

Let us assume now that $\sin(\varphi_2 - \varphi_1) \approx 10^{-2}$. Then we shall find the force (9):

$$\Delta F \approx 20 \text{ [dynes].} \tag{14}$$

From (10, 14) we find torque:

$$M \approx r \cdot \Delta F \approx 2 \text{ [dynes*cm].} \tag{15}$$

7. The Equations of Turbulent Flow

Let us return to formula (5.1):

$$\overline{F_{21}} = \frac{\varsigma\xi G m^2}{c^2 r^3} \left(\overline{v_1} \times \left(\overline{v_2} \times \overline{r}\right)\right)\left[dynes = \frac{g \cdot cm}{sec^2}\right]. \tag{1}$$

Similarly to p. 5 we find

$$\overline{\Delta F} = \vartheta \cdot \overline{\Delta f}, \tag{2}$$

where

$$\vartheta = \frac{\varsigma\xi G m^2}{c^2 r^3} \left[\frac{g}{cm^2}\right], \tag{3}$$

$$\overline{\Delta f} = \vartheta\left(\left(\overline{v_1} \times \left(\overline{v_2} \times \overline{r}\right)\right) - \left(\overline{v_2} \times \left(\overline{v_1} \times \overline{r}\right)\right)\right). \tag{4}$$

Taking into account (11b), we rewrite (3) as

$$\vartheta = \frac{\varsigma\xi G \rho^2 d^6}{c^2 r^3} \left[\frac{g}{cm^2}\right]. \tag{4a}$$

Further we shall denote the forces causing the turbulence, as T. In the Supplement is shown (see also Fig. 1), that if all forces lie in one plane, then (4) is equivalent to formula

$$T_y = \vartheta \cdot R_x \left(v_{2x}v_{1y} - v_{2y}v_{1x}\right), \tag{5}$$

where

T_y - is a force acting on the mass moving with speed v_2,

R_x - the distance between the masses centers.

Let the two adjacent molecule groups are located on the ox axis. We denote

$$R_x = dx, \tag{6a}$$

$$v_2 = v, \quad v_1 = v + dv. \tag{6в}$$

Then

$$T_y = \vartheta \cdot dx \left(v_x \left(v_y + dv_y\right) - v_y \left(v_x + dv_x\right)\right) \tag{7}$$

or

$$T_y = \vartheta \cdot dx \left(v_x dv_y - v_y dv_x\right) \tag{8}$$

Similarly, for a right coordinate system we have:

$$T_z = \vartheta \cdot dy \left(v_y dv_z - v_z dv_y\right) \tag{9}$$

$$T_x = \vartheta \cdot dz \left(v_z dv_x - v_x dv_z\right). \tag{10}$$

Let us consider an operator (further for shortness sake we shall call it turbulean)

$$\Omega(v) = \begin{vmatrix} v_z \dfrac{dv_x}{dz} - v_x \dfrac{dv_z}{dz} \\[2mm] v_x \dfrac{dv_y}{dx} - v_y \dfrac{dv_x}{dx} \\[2mm] v_y \dfrac{dv_z}{dy} - v_z \dfrac{dv_y}{dy} \end{vmatrix} \left[\dfrac{cm}{cek^2}\right]. \tag{11}$$

Example 1. We shall consider an ideal laminar flow in which $v_x \neq 0$, $v_y = 0$, $v_z = 0$. Apparently here $\Omega(v) = 0$, i.e. laminar flow cannot spontaneously become a turbulent flow.

According to (6a) we have
$$R = dx = dy = dz \tag{12}$$
From (10-12) follows the expression

$$T = R^2 \vartheta \cdot \Omega(v)\left[cm^2 \, \frac{g}{cm^2} \cdot \frac{cm}{sec^2} = \frac{g \cdot cm}{sec^2} = dynes \right]. \tag{13}$$

или

$$T = \vartheta_1 \cdot \Omega(v)[dynes], \tag{14}$$

где

$$\vartheta_1 = R^2 \vartheta = \frac{R^2 \varsigma\xi G \rho^2 d^6}{c^2 r^3}\,[g]. \tag{15}$$

The expression (14) defines a force acting on the group of molecules from the side of three adjacent molecule groups, located before the first group on the coordinate axes, if the differentials of the coordinates are equal to the distance between molecules (12). This force is acting on the volume of four molecule groups, i.e. on volume $4d^3$. So the force acting on a unit volume is

$$T_m = \rho_m \Omega(v)\left[\frac{dynes}{sm^3} = \frac{g}{sec^2 sm^2}\right], \tag{16}$$

where

$$\rho_m = \frac{\vartheta_1}{4d^3} = \frac{R^2 \varsigma\xi G \rho^2 d^3}{4c^2 r^3}\left[\frac{g}{cm^3}\right]$$

2 3

or

$$\rho_m = \frac{\varsigma \xi G \rho^2 d^8}{4c^2 r^3} \left[\frac{g}{cm^3} \right], \tag{17}$$

поскольку $R \approx d$.

Note for comparison, that in hydrodynamics equations, the dimension of mass force is $F_m \left[\dfrac{dynes}{g} = \dfrac{cm}{sec^2} \right]$, and the dimension of force acting on a unit volume. is $\rho F_m \left[\dfrac{dynes}{g} \dfrac{g}{sm^3} = \dfrac{dynes}{sm^3} = \dfrac{g}{sec^2 cm^2} \right]$. The dimension of force (16) is exactly the same. The coefficient (17) has the dimension of density and it can be called the turbulent density of a liquid.

> **Example 2.** Let us find the turbulent density ρ_m of water. We have:
> $$\rho = 1 \left[g/cm^3 \right] \quad d \approx 0.1 [cm], \quad c \approx 3 \cdot 10^{10} [cm/sec], \quad \varsigma = 2,$$
> $\xi \approx 10^{12}$. Let the diameter of the stream is $d \approx 0.1 [cm]$ and the distance between the streams is $r \approx 10^{-8} [cm]$. Then
> $$\rho_m = \frac{\varsigma \xi G \rho^2 d^8}{4c^2 r^3} = \frac{2 \cdot 10^{12} \cdot 7 \cdot 10^{-8} \cdot 10^{-8}}{4 \cdot \left(3 \cdot 10^{10}\right)^2 \left(10^{-8}\right)^3}$$
> or $\rho_m \approx 0.4 \left[\dfrac{г}{см^3} \right].$

The forces (16) may be included into Navier-Stokes equations. The Navier-Stokes equations complemented by such forces become the equations of hydrodynamics for turbulent flow.

The turbulean (11) by its structure is similar to the expression

$$(v \cdot \nabla)v = \begin{bmatrix} v_x \dfrac{\partial v_x}{\partial x} + v_y \dfrac{\partial v_x}{\partial y} + v_z \dfrac{\partial v_x}{\partial z} \\[2ex] v_x \dfrac{\partial v_y}{\partial x} + v_y \dfrac{\partial v_y}{\partial y} + v_z \dfrac{\partial v_y}{\partial z} \\[2ex] v_x \dfrac{\partial v_z}{\partial x} + v_y \dfrac{\partial v_z}{\partial y} + v_z \dfrac{\partial v_z}{\partial z} \end{bmatrix}, \qquad (18)$$

appearing in the Navier-Stokes equation. Thus for turbulent flows calculations one may use the known method for the Navier-Stokes equations solution, including the method presented in [6, 7].

The expression (18) arrears in Navier-Stokes solution with a factor ρ. Consequently, the turbulean (11) can influence the equation's solution if the coefficient (17) will be equal $\rho_m \approx \rho$.

Supplement
Let us consider an expression with vectors
$$\bar{f} = \left(a \times \left(b \times \bar{r}\right)\right). \qquad (1)$$
In a right Cartesian coordinate system this expression will look as follows:

$$\bar{f} = \begin{bmatrix} a_y\left(b_x r_y - b_y r_x\right) - a_z\left(b_z r_x - b_x r_z\right) \\[1.5ex] a_z\left(b_y r_z - b_z r_y\right) - a_x\left(b_x r_y - b_y r_x\right) \\[1.5ex] a_x\left(b_z r_x - b_x r_z\right) - a_y\left(b_y r_z - b_z r_y\right) \end{bmatrix}. \qquad (2)$$

Let us assume that the projections of these vectors on the z axis are equal to zero. Then

$$\bar{f} = \left(b_x r_y - b_y r_x\right) - a_x \begin{bmatrix} a_y \\ \\ 0 \end{bmatrix}. \qquad (2a)$$

Let us also assume that $r_y - 0$, i.e. $r = r_x$. Then

$$\bar{f} = rb_y \begin{bmatrix} -a_y \\ a_x \\ 0 \end{bmatrix}. \tag{3}$$

So, in the assumed conditions

$$\bar{f}_{ab} = \left(\bar{a} \times \left(\bar{b} \times \bar{r}\right)\right) = rb_y \begin{vmatrix} -a_y \\ a_x \end{vmatrix}. \tag{3a}$$

Similarly

$$\bar{f}_{ba} = \left(\bar{b} \times \left(\bar{a} \times \left(-\bar{r}\right)\right)\right) = -ra_y \begin{vmatrix} -b_y \\ b_x \end{vmatrix}.$$

We have

$$\overline{\Delta f} = \bar{f}_{ab} + \bar{f}_{ba} = r \left(\begin{vmatrix} 0 \\ a_x b_y - a_y b_x \end{vmatrix} \right) \tag{4}$$

or

$$\overline{\Delta f_y} = r\left(a_x b_y - a_y b_x\right) = rab\left(\cos\varphi_a \sin\varphi_b - \sin\varphi_a \cos\varphi_a\right), \tag{5}$$

where φ_a, φ_b - the angles of vectors a, b with the axis ox. Thus, vector $\overline{\Delta f}$ lies in the same plane where the initial vectors are located, this vector is directed along oy axis and has the value

$$\Delta f = rab\sin\left(\varphi_b - \varphi_a\right). \tag{6}$$

References

1. Khmelnik S.I. Experimental Clarification of Maxwell-similar Gravitation Equations. This issue.
2. Samokhvalov V.N. Papers in this issue.
3. Ivanov B.N. World of physical hydrodynamics. From the problems of turbulence to the physics of the cosmos. Ed. 2nd. - Moscow: Editorial URSS, 2010 (in Russian)
4. Zilberman G.E. Electricity and Magnetism, Moscow. "Science", 1970 (in Russian)
5. Wilner J.M. etc. Handbook of hydraulics and hydraulic drives, ed. "High School", 1976 (in Russian)

6. Khmelnik S.I. The existence and the search method for global solutions of Navier-Stokes equation. «The Papers of Independent Authors», Publisher «DNA», Israel, Printed in USA, Lulu Inc., catalogue 9748173, vol. 17, 2010, ISBN 978-0-557-88376-9

7. Khmelnik S.I. Navier-Stokes equations. On the existence and the search method for global solutions (second edition). Published by "MiC" - Mathematics in Computer Comp., printed in USA, printed in USA, Lulu Inc., ID 9976854, Israel, 2010, ISBN 978-1-4583-2400-9.

Series: **PHYSICS and ASTRONOMY**

Elkin I.A.

Model of Inertia and Gravitating Mass. Mechanism of Gravitation.

Annotation

It is not understanding, why physicists haven't paid attention to the elementary solution of the question about the formation of inertia mass and elementary explanation of gravity force till now. As soon as velocity-addition formula has been received by Einstein, it has become clear that derivative of total velocity of one of velocities gives two summands. At that it was clear that there can be the situation, when additional speeds would be for some object with plus and with minus at once. In this case the result for two results of total velocity can be averaged and it would be possible to receive the summand of concrete figure instead of zero result which depends on the direction of varying velocity. This improvidence of scholars causes that last one hundred years scholars have been working meaninglessly but not for the development of science in this sphere.

Contents

1. Mechanism of Gravitation
2. The crude estimate of masses
References

1. Mechanism of Gravitation

Till the field of gravity isn't uncovered experimentally: gravity waves are not uncovered as the carrier of interaction. We have only indirect proof - gravitation of electrically neutral objects. But it's clear that the proof of occurrence of some result isn't a proof of occurrence the source itself (from which we have a result). In this case we have a question: maybe there is no field of gravity and what can we offer instead of gravitational interaction? Because the interaction of electrically neutral objects really exists. What if we try to make the model of mass and the model of interaction of electrically neutral objects, without field of gravity?

What do we need for this:

1.1. We should use the force which is scientifically known – for example electric force.

1.2. We should receive the force of interaction forty orders less than the electric force directed to gravitation.

1.3. We should differentiate impulse in time to receive force. Impulse depends on the speed of movement. So, in the process of differentiation of impulse in time we come to the speed differentiation in time. If all the speeds are in one inertial system reference (ISR) all the addings to the speed occur in linear. And differential will be linear combination of speed derivative and adding. If the speeds for the reasons are in different (ISR) the addings to the speeds occur by addition formula of speeds. In this case, the process of differentiation will be more complicated function because we differentiate addition formula by speeds. We may receive the force, which we haven't observed earlier. For the realization of the given model at least two speeds in different frames should exist. And some addings to these speeds also should exist.

1.4. In order that the model works out it is necessary that there wouldn't occur forces, that are unknown in science, or forces that neutralize the received force in the offered interaction model.

1.5. We should be distracted from that we are in nonhomogeneity (but not in isotropic and homogeneous Universe) – Galaxy, planetary system and so on. That this nonhomogeneity is restrained by the constraint forces because a model of one of these forces is observed.

1.1. As far as we search the interaction changing the gravitation (in other words gravitation, correspondingly there is no potential for gravitation) there is also no gravitational changing of metrics when considering the local metric scales (changing of metrics from the nearest celestial bodies and other bodies). In other words the changing of metrics from the nearest bodies don't neutralize cosmological changing of metrics. As far as there is red shift we'll consider that there is cosmological divergence as an integral part of changing of cosmological metrics. We'll not discuss its origins and the causes of changing. For example, it is Friedmann-Robertson-Walker metric. Practically, the reasons of this "divergence" is not important yet, its only nessesary to know that this "divergence" really exists. As the type of addition formula of velocities isn't important – whether it is type of Einstein or there is some other type. Of course, we use Einstein type of formula because we have no alternative variant now.

In this case, well-known fact: when considering the local domains there is vanishingly small divergence **U** material points, connected with cosmological divergence. And so there are no local variation of metrics because of the proximate mass (because there is no gravitation) cosmological divergence is a single divergence because of the changing of metrics. Note, that – accelerated divergence.

That's why we'd better consider this divergence at quite long range not to separate out some concrete point. We consider that given material points are deleted from their original position (in the process of divergence) at a speed of **u**. I'd like to make arrangement about the using of several velocities for one material point at one moment, for the physicists not to ask questions about the strict formulating of determination. Separately these velocities are pointless, resultant is necessary, to find it out firstly we take the velocities separately. To use this electric force instead of gravitation force we should use two neutral objects consisting of charged particles. The participle of one object interact with the participle of the other object on the principle of superimposition, interaction set the speed to the participle (in fact they differ). We'll consider they to be equal in absolute magnitude as far as this very result is interesting (and it doesn't change the similarity of verbal proves). After that the interties of these objects compensate these interactions, all this is difficult and takes a lot of time. Without breaking the line of reasoning I'll consider one object as a material point **A** with electric charge. The other object will be a material point **D**, consisting of two material points B and C with different charges which affect **A**. In fact **A** and **D** are electrically neutral to each other. Arising in the process of interaction speeds and forces will belong only to **A,** because only resulting force affecting **A**, is interesting. Of course, these forces and speeds are for different participles. We'll not break the generality, because in total we are interested in one material point and after all common compensations it's resulting speed and force affect it. Consideration of material point comes to the consideration of macro-systems and release from the consideration of quantum mechanical effect.

1.2. I'll recollect that we consider the divergence of material points, not connected with inertial motion but with the changing of the metrics. At that we consider two divergence material points in some distance. Each material point is moving away from its original position with some speed. We'll consider some fixed ISR1, connected with **D'**. Just yet without the determination of the origin of speeds: we have low speed **u** in ISR2 (relative to ISR1), connected with **D**, where **D** is that very material point **D'** at the next moment of time. And it moves relative to the fixed ISR1.

This material point **A** may have adding to speed **+V** or separately **-V** (because of electric interaction with **B** or separately with **C**). If we now consider fixed **D**, and connected with it ISR, due to the addition formula of speed from ISR we'll receive the speed of movement **A** relative to **D** for every variant of electric interaction:

$$w = \frac{(v+u)}{1+\dfrac{vu}{c^2}}.$$

Now in formula for force (force expression through impulse) we have additional terms:

$$F = \frac{dP}{dt},$$ where $P = P(w)$ is the dependence of impulse on speed –

only the variant of change of speed in [1]. Then:

$$F = \frac{dP}{dt} = A\frac{dw}{dt},$$

where $A = \dfrac{m}{(1-\dfrac{w^2}{c^2})^{\frac{2}{3}}}$ (this is from that very source), and $\dfrac{dw}{dt}$ is the time

derivative of the formula of speed addition.

We'll put it into the formula of force:

$$F = A[\frac{\dfrac{dv}{dt}}{(1+\dfrac{vu}{c^2})} - \frac{u+v}{(1+\dfrac{uv}{c^2})^2}\frac{dv}{dt}\frac{u}{c^2}] = f\frac{1-\dfrac{u^2}{c^2}}{(1+\dfrac{vu}{c^2})^2},$$

where $f = A\dfrac{dv}{dt}$ – expression for force (from that very source), in the

process of changing the velocity in magnitude. We'll take positive direction for **f** and **V** will be the direction for removal and we'll find average force between the repulsion force **A** and force of gravity **A**, (correspondingly we use speeds: **-V** and **+V**):

$$F_{Average} = \frac{1}{2} f (1 - \frac{u^2}{c^2})(\frac{1}{(1+\frac{uv}{c^2})^2} - \frac{1}{(1-\frac{uv}{c^2})^2}) = -f(1-\frac{u^2}{c^2})\frac{\frac{2uv}{c^2}}{(1-(\frac{uv}{c^2})^2)^2} \quad (1)$$

In this formula **f,V** are absolute magnitudes, clearly, that $F_{Avg} \square f$, because $u, v, \frac{1}{c^2}$ linear enter into a formula. It is clear that with quite low speed value of the required correlation in 40 times is received. At that the force is always negative, so there is always gravitation there.

I'd like to mark specially that speed **+u** is coming, as divergence with one sign **+**. But if it would be necessary to consider formula (1) for some speed at approaching, so speed **-V** at that formula (1) would be the formula of repulsion force.

1.3. It is necessary to find two frames of reference and add the moving from the known interaction to their moving. It's clear that considered earlier ISR2 isn't that one frame of reference, in which charged material points have been interacting. That's why the speed of divergence of ISR2 and speed received in the result of electric interaction should be combined due to the addition formula of speeds.

We have at once a question as for the interaction of accelerating, for example, rocket with accelerating Earth (we mean accelerating when moving around the Sun). In this case the accelerating of Earth and the rocket are in one frame of reference though this frame of reference is accelerating. In that time the cosmological divergence speeds up every participle without its dependence on the other participle.

Formula (1) gives an opportunity to determine the gravitating mass. Now we may try to make a model of formation of inertial mass. If we consider all the participles of the Universe (lets' consider it to be isotropic) **D** will interact with each of them. In other words every material point of the Universe gravitates to itself **D**, if we consider **D** to be static. At that they will gravitate equally in all the directions because the Universe is identical and there are no reasons to gravitate **D** more in some concrete direction. If we speed **D** up, speed of divergence in accordance with moving with the material points of the Universe will decrease at the moment of acceleration and the speed of divergence in the opposite direction will increase. Correspondingly, due to the formula (1) the gravitation in direction of moving will decrease and in the other direction it will increase. Here are inertial and gravitating mass. No doubts, in any cases the force effective on **D is** proportional to charges that this material point includes, in other words, the masses are proportional.

2. The crude estimate of masses

2.1. It's clear that gravitating mass should be received from formula (1). There are 2n of charges with average charges **q** at the material point. These points interact with **2N** of charges and average charge **q'**. If we take off multipliers uniquely defined we'd receive the following formula (1)

$$F = -f \frac{2uv}{c^2} = -k \frac{qnq'N}{r^2} \frac{2uv}{c^2} \tag{2}$$

it's clear that there is a formula that could be referred to mass **D**,

$$m_D = Mqn \tag{3}$$

where M is some multiplier. In this case we may refer to the gravitation constant:

$$\Gamma = \frac{k}{M^2} \frac{2uv}{c^2} \tag{4}$$

2.2. In the case of inertial mass we'll take off multipliers uniquely defined by the same way. But instead of **V** we should take its changing (let's consider that only decreasing and increasing relate to the equal value, though it isn't so) connected with acceleration of object **D**. This would change **dV**, for the short period of time $_0$.

We should notice that **V** also has been determined by this short period of time. Let's consider that there is the number of charged participles in the Universe **2P** and their average charge is **Q** and they are in average distance **R**.

At this case:

$$M_{inert(D)} = \frac{F_{universe}}{a} = k \frac{qnQP}{R^2} \frac{2u}{c^2} \tag{5}$$

(Explanation: inertial mass of body **D=** force of gravity of the Universe divided into acceleration conventional sign).

We have just to find these electrical components of the objects to compare inertial and gravitating masses. And will it be enough this description by quarks for exhaustible setting this model of mass is a question.

References

1. L.D. Landau and E.M. Lifshitz, «Fiel Theory», P.2, Moscow chief editorial board of physico-mathematical literature, 1967г., printed sheets 28,75., Page. 43.
2. January 10, 2013. Igor Elkin, http://elkin-igor.livejournal.com/

Series: PHYSICS and ASTRONOMY

Elkin I.A.

The causes of the contraction in the direction of motion. Metric of Universe. The Causes of Red Shift.

Annotation

Poincare and Einstein supposed that it is practically impossible to determine one-way speed of light, that's why speed of light in different ways may differ. They also supposed that till there is no the experiment which would depend on the value of one-way speed of light, it is possible to consider that all one-way speeds of light are equal to two-way speed of light. It is offered the explanation of the experiment connected with "Red shift" with the aid of one-way speed of light, therefore it is shown the dependence on magnitude of one-way speed of light. Mechanics is set up based on the unidirectional speeds of light. For example, the experiment of Michelson-Morley is explained with the aid of this mechanics. The cause of the inception of Fitzgerald contraction is explained.

Contents

1. The Red Shift

Introduction

For the beginning we'd most like to remind, how Einstein came to the idea of the inalterability of the speed of light in any direction. The secret is in the accepted clock synchronization, which as it was considered makes undetectable difference of the one-way speeds of light. Inasmuch as the synchronization converts to the moving from some point (with time on the clock "0") of the synchronized light signal and with transmission of this signal into another synchronized point by the setting in this point of time on the clock. Time on the clock in this another point is fixed equal to the half of the time of signal transmission of the way to this another point (from starting point) and by the way of the getting back returned signal into the initial point. That's why it isn't important with what one-way speed the light-signal moved on either side, just middle speed is important. The procedure of length measurement, set up on the basis of this clock synchronization, made possible a lot of experiments, where different one-way speeds of light are possible and we may trace them to the analogical experiments with two-way speeds of light. Einstein suggested that if experiments lead to the two-way speeds of light and there is no dependence on the one-way speeds of light, there is no sense to consider them. That time a postulate was taken which tells that all one-way speeds of light are equal and they are equal to two-way speed of light.

Let's try to receive with the aid of different one-way speeds of light ("to observer" and "from observer") generally known experimental fact, precisely: emission spectrum comes shifted into the red side from the distant galaxy, and as this galaxy is on the longer distance, as the shift becomes bigger.

1.1. The Red Shift – Real Cause [1]

Let's consider our galaxy and some distant galaxy. We take it that two-way speeds of light are unchangeable everywhere, it means that emission in these two galaxies in relation of two-way speed of light is unchangeable and it has one wave-length, we mark it as: L. It is really so, because a metre is mensurated with two-way speed of light. That's why as this speed of light will be changed, a metre will be also changed and it'll be impossible to fix this change.

Now we consider that the light that comes to our direction from the side of distant galaxy at speeds of $c^- < c$ where c — is two-way speed of light.

Difference in time of arrival of fore and back wave front $\Delta t = \dfrac{\Delta L}{c^-}$,

but as it is known on the Earth we conceive this time as coming by light-signal some way with two-way speed of light. Or else at short distances one-way speeds of light less differ from the two-way speeds of light than at long distances. That's why, we on the Earth, consider

that the wave-length of the light-signal is $\Delta l = c\Delta t = \Delta L(\dfrac{c}{c^-})$. In other

words wave-length was increased. Correspondingly the frequency was decreased and we received "The Red Shift"

The data of these researches may help estimate one-way speed in the distance of one megaparsec. By estimate of cosmologists, it's nearly:

$c^- = c - 70\,\text{km/sec}$

Now we'll mark $k = \dfrac{c^-}{c^+}$

Than we'll find the correlations on the way **h**:

We should mention that these formulas are taken for the estimation of one-way speeds of light, more proximate formulas we may use after numerical calculations of metrics — this is in item 2 part 2.

$T_1 = \dfrac{h}{c^-}$, $T_2 = \dfrac{h}{c^+}$ at this case mid-speed on the way **2h** is:

$$c = \frac{2h}{T_1 + T_2} = \frac{2c^-c^+}{c^+ + c^-}$$

or

$$c = c^- \frac{2}{1+k} = c^+ \frac{2k}{1+k} \qquad\qquad (1)$$

or

$$(1+k) = 2\frac{c^-}{c^+} = 2(1 - \frac{70}{300000})$$

or roughly:

$k = 1 - 0,0005$

We haven't received yet the dependence of shift on the distance. It is deducted a bit more complicated. Let's think back that clock synchronization initiated by Poincare and then transformed by Einstein uses two-way speed of light as synchronizing signal. Though Einstein wrote that it is possible to use any signal, but the speed should be fixed and measured by some way. That's why at any case clock setting in total comes into synchronization by the light signal with two-way speed.

The dependence of wave-length of incoming signal on the one-way speed of light, received so elementary, put in doubt that one-way speeds of light can't influence on the result of any experiment and it is possible at any case to put in two-way speed of light. As far as this hypothesis may be accompanied by serious mistakes and loss of real causes of events is also possible, - that is observed in Physics. As far as the setting of "synchronism" in two points is fixed, so:

Synchronization: let's consider that with light signal output (or analogical to it signal) zero is set and with coming of light signal zero is also set in the correspondent point – for the given process. The moving in some concrete direction is considered as the given process. At this case there is a necessity to consider all the processes separately from each other.

This synchronization changes the concept of simultaneity, and also type of time coordinate. And if the time coordinate is changed the type of interval is also changed, and type of metric tensor is changed too. It may mean that space metrics is also changed. We'll try to investigate this question.

1.2. New Time Coordinate. Metrics of the Universe

We consider it in the context of new synchronization. If we mark material point at the beginning of the coordinates – M_0, and the other arbitrary point - M_1, time coordinate of the point M_1 - t_{M1} would depend on the direction of the signal by this way:

1) $t_{M_1} = t_{M_0} - \dfrac{R^+}{c^+}$ - in the case of sending of the signal from M_0 to M_1.

2) $t_{M_1} = -t_{M_0} + \dfrac{R^-}{c^-}$ - in the case of sending of the signal M_1 to M_0.

Where c^+ and c^- are correspondingly the speeds of light in the concrete side (we'll mark it), t_{M0} is the time coordinate of the point M_0.

I'll explain a bit: it's understandable that in 1) from the usual consideration of time at the point M_1 time for overcoming with light the distance between the points is just taken off.

It is analogically in 2), just in this case usual for us time is at the point M_0 is taken with minus, as far as in the case of slower synchronizing signal than light we receive information about the object from more distant past.

R^+ and R^- – Euclidean distance (length) between the considered points, in units of speeds c^+ and c^- respectively.

R is the length, in units of two-way speed of light **c**.

For simplicity we'll mark $t_{M1} = q_x$, $t_{M0} = q$. Then:

1) $q_x = q - \dfrac{R^+}{c^+}$ moving from zero.

2) $q_x = -q + \dfrac{R^-}{c^-}$ moving to zero.

Let's now consider moving to zero. Let's admit there is a dependence of one-way speed of light on **X**-coordinate. We'll search for the dependence like this:

$$\frac{c^-}{c} = f = f(x^-) \tag{2}$$

X-coordinate is defined in two-way speed of light, coordinate x^- is defined in two-way speed of light for approximating.

For Minkowski space in two-way speeds of light the next interval is known:

$$dS^2 = c^2 dt^2 - dx^2 \tag{3}$$

$$dx = \frac{c\, c^-}{c\, c^-} dx = \left(\frac{c^- dx}{c}\right)\frac{c}{c^-} = \frac{c}{c^-} dx^- = \frac{dx^-}{f} \tag{4}$$

In terms of $q_x = -q + \dfrac{x^-}{c}$ for event space (which is corresponded to Minkowski space) letter **q** means time at the observing point, in other words **t**, so, it means that in the coordinates of one-way speed of light time:

$$t = \frac{x^-}{c} - q_x = \frac{1}{f}\frac{x^-}{c} - q_x$$

or

$$dt = -\frac{f'}{f^2}dx^-\frac{x^-}{c} + \frac{1}{f}\frac{dx^-}{c} - dq_x \qquad (5)$$

substitute in (3) formulae (4) and (5). And we consider just space part of interval:

$$dS^2 = c^2(\frac{1}{f}\frac{dx^-}{c})^2(1-\frac{f'}{f}x^-)^2 - (\frac{dx^-}{f})^2$$

$$dS^2 = \frac{1}{f^2}(dx^-)^2\frac{f'}{f^2}(x^-)[(x^-f'-2f] \qquad (6)$$

(factor out and product of difference of squares)

We'll seek the solution using the method of Occam's razor. It's understandable that the most simple type of interval is at this case: $f' = A = const$, I'll remark, that then formula:

$f = Ax^- + B$, where $B = const$

Can be easily taken in the form of a bit changed Hubble's law, which is easily explained for the case of different one-way speeds of light. This law works in case that:

$A = -\dfrac{H}{c^2}$ and $B = 1$, where **H** – Hubble constant. In other words one-way speed of light is decreased when the distance is increased. Or

$$f = 1 - \frac{H}{c^2}x^-$$

Square brackets in formula of interval will be simplified by this way:

$$[(x^-f'-2f] = -2\frac{H}{c^2}x^- - 2 + \frac{H}{c^2}x^- = -(2-\frac{H}{c^2}x^-)$$

The interval will contain just space components, in other words it will be metrics of space on this direction

$$dr^2 = \frac{H}{c^2}(x^-)\frac{(2-\frac{H}{c^2}x^-)(dx^-)^2}{[1-\frac{H}{c^2}(x^-)]^4} \qquad (7)$$

It is understandable that the *Pythagorean theorem* doesn't work, in other words one-sided metrics in unilateral coordinates is non-Euclidean. This metrics is of the form of space which is described by Lobachevskian geometry.

Formula (1) proves that, one-way speeds of light are connected with two-way speed of light asymmetrically. That's why metrics will be different on the direction "from the observer" and "to the observer". The non-Euclideanism of metrics explains the changes of one-way speed of light. As far as metre in the space with non-Euclidean, metrics isn't invariant. In other words, subpaths of light signal which are equal for the space with non-Euclidean metrics will be of different length, because length – is Euclidean metric. And light, of course, when moving may use just the characteristics of the space (in non-Euclidean metric), but not foolish theories of the scientific men, who require fixed speed of light in the metrics of another's space.

2. The Causes of the Contraction in the Direction of Motion

2.1. Michelson-Morley Experiment [2]

In the result of getting new clock synchronization and researching of different one-way speeds of light "to the observer" and "from the observer" space metrics of Universe was received. Formulae are elementary and after simplification they easily transfer into short and understandable ones. That's why their original long and boring states shouldn't distract.

I'll recollect shortly:

The following formulae were received:

$$dt = -\frac{f'}{f^2}(dx^-)\frac{x^-}{c} + \frac{dx^-}{fc} - dq_x \tag{8}$$

$$dx = \frac{cc^-}{cc^-}dx = \frac{c^-dx}{c}\frac{c}{c^-} = \frac{dx^-}{f}, \tag{9}$$

where

c^- - one-way speed of light to the "observer",

$\dfrac{c^-}{c} = f = f(x^-)$ - the dependence of one-way speed of light on

x^--coordinate.

x -coordinate is defined in two-way speed of light,

coordinate x^- is defined in one-way speed of light for approximating,

q — means time at the point of observer,

q_x - means time at the observed point.

Formula of interval:

$$dS^2 = c^2dt^2 - dx^2 \tag{10}$$

If we put (8), (9) in (10), we'll receive interval in one-way speeds:

$$dS^2 = c^2dq_x^2 - c^2(\frac{1}{fc} - \frac{f'}{f^2}\frac{x^-}{c})dx^- dq_x + \frac{1}{f^2}(dx^-)^2\frac{f'}{f^2}(x^-)[(x^-)f' - 2f]$$

or

$$dS^2 = cdq_x^2[c - \frac{dx}{dq_x}\frac{1}{f}(1 - \frac{f'}{f}(x^-)] + (dx^-)^2\frac{f'}{f^4}(x^-)[(x^-)f' - 2f] \tag{11}$$

With regard:

$$f = 1 - \frac{H}{c^2}(x^-) \tag{12}$$

$$dS^2 = cdq_x^2(c - \frac{dx}{dq_x}\frac{1}{[1-\frac{H}{c^2}(x^-)]^2}) + \frac{H}{c^2}(x^-)\frac{(dx^-)^2(2-\frac{H}{c^2}(x^-))}{[1-\frac{H}{c^2}(x^-)]^4} \tag{13}$$

This long formula just proves that for long distances space metrics is Lobachevsky metrics.

It's not difficult to check that for the speed of light to the$_+$observer we'll receive analogical formula (just space coordinate will be x and it is defined with the speed of light for moving off).

We clarified with recession of galaxies in the first article. (reference in the beginning). It's understandable that only the magnitude of one-way speed of light impacts on the red shift. And there is no universal red shift. Now Michelson-Morley experiment and physical meaning of the length contraction in the direction of motion are interesting. In fact STR doesn't give physical causes and physical meaning of this contraction and it is not understandable what for fields (and space) should reduce their length. We can explain it even with the aid of the magnitude from other ISO as far as the procedure of the length determination by Einstein wouldn't afford it.

The interval becomes easier in short distances:

$$dS^2 = cdq_x^2(c - \frac{dx^-}{dq_x}) + a*(x^-)*(dx^-)^2 \qquad (14)$$

where

$$a = \frac{H}{c^2}$$

In this case, the space part of the interval will give us space metrics (for the simplicity we'll use instead of x^- - $\boldsymbol{x}$):

$dr^2 = ardx^2$ or

a - is scale number, it's understandable, that further it can be counted as equal to unity element.

$$dr = \sqrt{x}dx \qquad (15)$$

This formula shows what metrics should be accounted in short distances. Ordinary scale number takes into account numbers of dimensions.

The formula itself means that metrics, in other words, the distance between two points is equal to

$\sqrt{x}$ numbers of unit elements of equal open intervals of coordinates or just the same — the unit of measurement of coordinate of one-way speed of light. Under condition, that one point is at the

beginning of the coordinates with the coordinate   , and the other point has the **x** - coordinate.

Of course, we'll not follow verbal proof (all the subpaths of light signal in coordinates, measured by light are described in thousands of works) and we'll write at once formula with light clock? Moving from

the observer with a speed   . Coordinate is measured with the corresponding speed of light, in this case, we'll take into **account formula (8) for each interval** (clock length **L**):

$$\sqrt{ct_1} = \sqrt{Ut_1} + \sqrt{L}$$

It is understandable, that all the meanings are taken in absolute magnitude or

$$t_1 = \frac{L}{(\sqrt{c} - \sqrt{V})^2}, \text{ by analogy with motion into another direction:}$$

$$t_2 = \frac{L}{(\sqrt{c} + \sqrt{V})^2}, \text{ total time of the motion of the signal:}$$

$$t = t_1 + t_2 = \frac{2L}{(c-U)} \frac{(c+U)}{(c-U)} \tag{16}$$

Now by analogy we'll consider known formula with cross collocating of light clock

$$\sqrt{L^2} = \sqrt{(ct_3)^2} - \sqrt{(Ut_3)^2}$$

So, we receive total time:

$$t = 2t_3 = \frac{2L}{(c-U)} \tag{17}$$

We can see, that total time formula (16) and formula (17) differs by multiplier:

$$N = \frac{(c+U)}{(c-U)} \tag{18}$$

We can begin to lie as in STR, that this is a feature of space and it is reduced by unknown ways, and L is reduced with it too, or we may analyze what the time part of the interval can give us, because we consider time in motion.

It is understandable that interval (14) in short distances as for

cosmological measures, and at steady speed   of the moving of the mirror (moving of ISO) time part of the interval will be:

$dq^2 = cdq_x^2(c-U)$, what gives (for approximating):

$$dq_- = dq_x\sqrt{c(c-U)} \qquad (19)$$

It's understandable that the same time interval with the speed of light for deleting gives:

$$dq_+ = dq_x\sqrt{c(c+U)} \qquad (20)$$

It's also clear that the motion of light signal and ISO are taken to consideration in each case that's why for the experiment with light clock (or the same one with the experiment of Michelson-Morley), the necessary for us intervals will look like:

$$dq_1 = dq_x\sqrt{c(c+U)}$$

$$dq_x = dq_2\sqrt{c(c-U)} \qquad (21)$$

$$dq_2 = dq_x\frac{1}{\sqrt{c(c-U)}} \qquad (22)$$

And formula, describing time of absolute signal transmission, will be (I shan't describe it, because it is easy for understanding):

$$dq_{sum} = dq_x\frac{\sqrt{c(c+U)}}{\sqrt{c(c-U)}} = dq_x\frac{\sqrt{c+U}}{\sqrt{c-U}} \qquad (23)$$

So, it appears that there is the increasing of the unit of time in the direction of the motion in the moving ISO.

And if there is change like that and there is the definition of metre, connected with the unit of time. It is obtained that metre in moving ISO also increases in corresponded number of times, that's why our light clock obtains less of these metres (and the arm in Michelson–Morley experiment respectively). So:

$$L' = L\frac{\sqrt{c-U}}{\sqrt{c+U}} \qquad (24)$$

That's why if in the formula (16) instead of L we'll write L'and put the meaning by formula (24), multiplier n will be left:

$$n = \frac{\sqrt{c+U}}{\sqrt{c-U}} \qquad (25)$$

The value of transmission time with light signal over the way in fore-and-aft and cross direction differ on this multiplier. And this multiplier is explained with the difference of time metrics in fore-and-aft and cross direction. In other words without contrive, causative «space contraction» we can easily explain the difference of transmission time with light-signal of fore-and-aft and cross direction. And short explanation:

In ISO moving at speed **U** relatively stationary observer, we can observe anisotropy of timing metrics in the direction of moving. Because stationary observer is going to observe in moving ISO. And only in Einstein variant of the course of events location of observer doesn't matter, in our variant we should take into account this location.

Anisotropy of metric units occurs due to anisotropy, timing metrics in this ISO in the context of stationary observer (because metre is observed by the speed of light and time. These both factors clean the difference in transmission time with the signal fore-and-aft and cross direction in light clock (and correspondingly in Michelson–Morley experiment).

2.2. Correlation of one-way and two-way speed of light

I get back to the question of the formula for the middle speed of light. It is important because in case of determination the measurement of the way of light signals with the aid of two-way speed of light there is some delimitation below the measurements of the one-way speeds. This delimitation contains the formula:

$$c_{cp} = \frac{2c^+c^-}{c^+ + c^-}$$

As far as the delimitation is below in the measurement of one-way speed of light it couldn't be the arbitrary relation of wave-length (due to the given theory) in the emissions of distant and close galaxies. But these very correlations are observed.

Practically we should arise from the examination of the middle speed and we should estimate the distances in one-way speeds as it was doing for Michelson–Morley experiment:

$$S_+ = \sqrt{x^+} = \sqrt{c^+ dt}$$

$$S_- = \sqrt{x^-} = \sqrt{c^- dt}$$

Next,

$$c = \frac{S_+ + S_-}{\dfrac{S_+}{c^+} + \dfrac{S_-}{c^-}} = \sqrt{c^+ c^-} \tag{26}$$

- this formula has no delimitations below for the one-way speeds of light and the correlations of wave-length is not fixed. Consequently there are no discrepancies of the experiment and theoretical hypothesis now [3].

References

1. http://en.wikipedia.org/wiki/Redshift
2. http://en.wikipedia.org/wiki/Michelson%96Morley_experiment
3. http://elkin-igor.livejournal.com/

Series: **PHYSICS and ASTRONOMY**

Alexsey Katyshev

The unified field and interaction

Annotation

This work gives a description of the unified field which at the interaction with the matter changes its qualitative characteristics and develops the feedback: field-matter-field. The formula is derived that relates the gravitation and energy background of the system by the movement of its parts and the particle spin.

Contents

The picture of the world allows us to make an assumption that the matter distributed in the space is an expression for the properties of the field, which in its motion forms the matter as substance, giving it all the physical properties and the symmetry, while it (the field) itself is matter. It means that in the space matter forms matter. And so we may talk about the discreteness of Nature, about the conditionality and priority of the relations between the Nature's own elements. About the levels of relations , where the large quantitative characteristics of the lower levels (altogether 4 levels) are considerably reduced, We can say that all the relationships occur only by the constants and only in proportion to the constants. So the mathematics is prior, and not vice versa. The Nature has a quantitative side in the form of the elements of field and of matter (particles). It is the sole material component. There is a qualitative side, in the form of bringing a certain quantity of motion as a result of interaction. I.e., the primary mechanism of the Nature existence causes a rigid constructivity and self-organization.

1. The Original Mechanism of Nature Formation

Let us assume that there exists one (we do not talk here of its further inner structure) minimal particle (element) of the field – let us call it graviton (a code name). In its motion the graviton transfers a certain regular momentum to the interacting matter. And not only in its motion, but also at sufficient closeness. That means that graviton possess a real area of repulsion. The gravitational field, and so also the gravitons, uniformly permeating the space, present a quantum, but not always equal flow from all directions. And so any, arbitrary small or arbitrary large (up to a certain size) area of the space is crossed by flows of gravitons And if in this area there is no concentrated matter, then the gravitational field, and so also the gravitons, are crossing this space without interaction. If in a certain space area there exists concentrated matter, then the gravitons, crossing this space and penetrating into the area of concentrated matter, keep this matter from disintegration. In the case of stable particles the field may keep only a certain area of concentrated matter in the space.

The parameters of the field are: 10^{40} simultaneously in 1 m^3. approximately 18000 directions. Speed C.

The motion of particles forms the field. The fact is that a particle "does not know" where to move in the next moment. When a particle moves the time of actual interaction (between particle and graviton) in the direction parallel to the speed vector is maximal, for graviton, beside its own energy, has a real area of repulsion that is by an order of magnitude larger than the graviton itself. And decreases to 90 grades, where it is balanced by opposite quanta. And opposite the speed vector it is the contrary – the time is maximal and increases toward 90 grades. As it is and energy process, then if there is no other interactions and gravitation (conditionally), then the particle moves in a straight line. There appears a purposeful force that stipulate the motion. That is – motion creates motion.

The conclusion: the trajectory of uniformly moving particle consists of equal and repeating intervals within which it is moving irregularly. On low speeds the amplitude of irregular movement is small. On relativistic speeds it increases to such an extent, that we can talk about a particle as a wave, but only mathematically. There appears an indetermination of impulse within each interval. Indetermination of impulse within interval causes diffractional picture. The probability of approaching to the slit of collimator causes probability of deviation. The speed is discrete, especially on high speeds. I.e. it may be 0.9,........845. There is no intermediate speed.

Gravitational field forms also particle spin. The particle spins in two planes. Always by speed vector, where lineal sped of spin-up is equal to C. And perpendicularly to the speed vector, where the initial linear speed is 1/137C and depends on the energy of the system where it is located (this is one of gravitation elements).

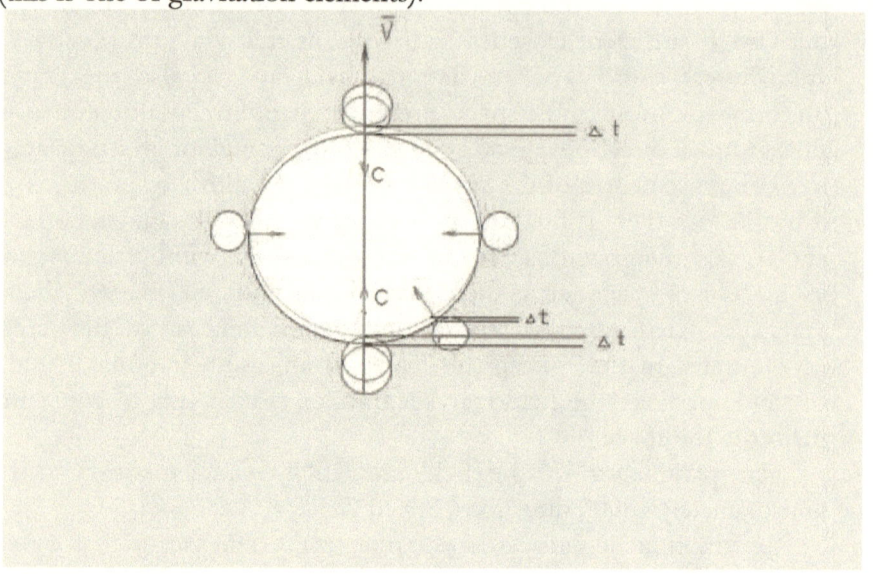

Particle in motion, in addition to absorbing the gravitons, emits them. The mechanism of emitting is spin. The frequency of emitted gravitons is maximal by their speed vector, and it decreases up to the opposite direction. When the speed grows, grows also the fields frequency by the vector, and vice versa. But the particle itself does not change its frequency, but there is redistribution of the field in the space (primitively it is so – ejection, advanced – another ejection). But for any speed the field frequency, perpendicular to speed, remains the same. Redistribution of the field's elements forms the background of the big system, where all its parts remain at rest related to the motion of the system. I.e. for them the field frequency of field remains the same for the system motion, as for full stop of the system. The same is true also for connected systems, such as galaxies and star systems.

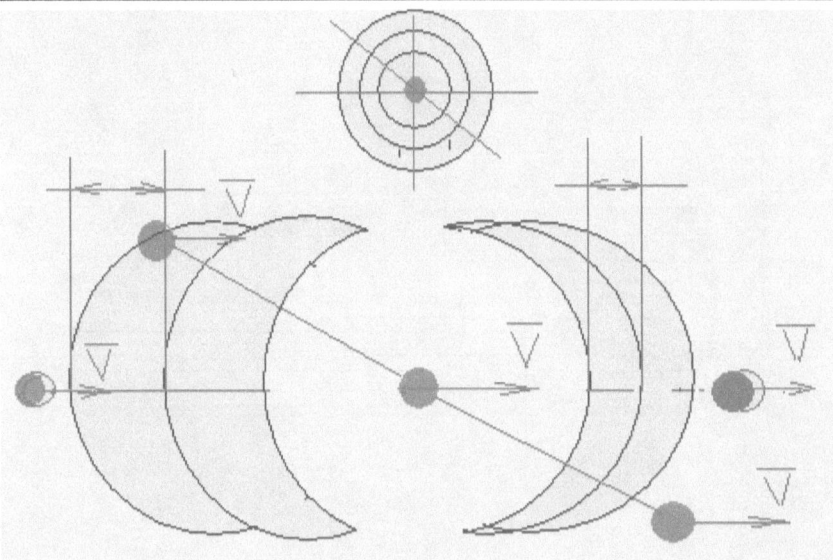

The particles emit gravitons by tangential to the plane of maximal rotation (where the speed is C). They are emitted simultaneously. A turn. Again emitting. So one may say that the emittion is discrete and distributed in the space. So there is a following picture: The field's redistribution for the moving particles remains the same as before. But along the vector and opposite to the vector of speed there appears a formation of particles like a spiral (spring) Along the vector the spiral is compressed and its energy properties grow. Against – vice versa. This is exactly the so-called photon. A particle always emits photon. Another thing – what a photon. For connected electrons in an atom the photon is "smeared", as the electron has a frequently changing direction. When the particle is moving, grows the frequency of the field by vector, but the mass doesn't grow, and the total energy grows. I.e. –energy of particle + energy of field.

In the usual state the total emittion of gravitons is distributed symmetrically in the volume, and so the total impulse is equal to zero.

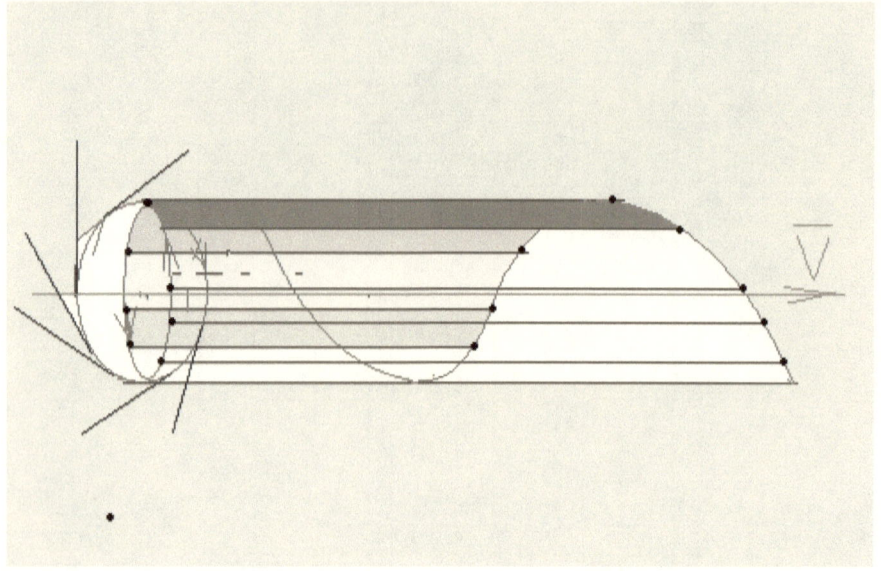

The energy background of the system is also determined by the motion.

I have explained before what is meant by photon. A particle emitting gravitons, rotates in two planes – along the speed vector – lineal speed C; and perpendicularly, where the linear speed is 1/137C. In order for the system to be in equilibrium, the number of absorbed and ejected gravitons should be equal. This is what the perpendicular component of spin is "dealing" with. Gravitons are emitted by the line of maximal speed. Ejected by tangential. It is easy to see that the frequency of ejection in this direction is

$$\omega 0 = \frac{Vs}{2\pi * Re},$$

where Vs – is the linear speed of the spin's perpendicular component, Re – the particle radius.

But the frequency here implies the number of consequent elements of the field in a time unit. The ejection parallel to the speed vector forms a stable formation photon.

As concerning Vs, we must say that Vs = 1/137C may be only in ideal, i.e. when the particle is not surrounded by other particles. When the surrounding exists, then the external field is partially overlapped and

Vs is reduced proportionally to k, where k is a number, by which the limiting number of the field's elements decreases in one taxonomic act (one field's element in the volume). $137^2 - k$.

A sole gravitation field as such doesn't exist. There are unified interactions, in which, due to the fact that the field of gravitons forms our matter and creates new qualities – particles (electrons, protons), they in their turn emit gravitons and create new qualities already in the field of gravitons. New qualities create various interactions of one field. Of course, it is a somewhat mechanistic approach to the model I have outlined here, but now this is the only way to do it now. From the above we may come to the conclusion, that based on a true initial background, where graviton is the smallest particle of the field on our quantum level, it, having a real repulsion area and obeying to sub quantum motion, forms the matter. And, based on interaction with the matter, forms material systems and physical laws related to motion.

2. asic Characteristics of Matter

The inanimate matter has three basic characteristics: motion, interactions and the process probabilities. Space is the volume created by a field having in ideal 137^2 directions depending on the distribution of matter in it. The field determines the symmetry of space, stability and symmetry of particles and their formations. The matter in motion comprises asymmetry of interactions. Stability ensures the conservation laws. Probability is due to the quantitative characteristics of the field. There ins no infinity and singularity in the Nature, and every feature is manifested as a process. Every motion occurs under the effect of a force. A graviton has a real repulsion area, and so the time of interaction is not uniform, but it is always an interval. Interaction time interval against a speed vector is the smallest and increases up to perpendicular, where it is equal to the interval of zero motion speed. Then it grows to the maximal by speed vector. (One can say – the instanteous speed). On the time interval depends the magnitude of energy interaction, and this value is the largest by the motion speed vector. And if we take the difference of energy interactions, it will be the largest by speed vector, and this is why the particle "knows" where to move in the next moment.

The resulting vector of forces is directed along the speed vector. These time intervals on quantum level represent time for the inanimate Nature, and it differs from the time for the Mind.

2.-L---------1.---V.L1-L2----L-.2

where

2 - is an element of field, vector of speed to a particle,

1 - a particle,

L- the radius or the length of the real repulsion area,

V- the speed of a particle with vector with direction V,

L1 – the length on which the particle interacts with the field along the speed vector,

L2 = against the vector.

Thus

$$D = V / C, \quad L1 = \frac{L}{1-D}, \quad L2 = \frac{L}{1+D},$$

$$L1 - L2 = L3, \quad L3 = \frac{2LD}{1-D^2}.$$

Time of interaction along vector $t1 = \dfrac{L}{C*(1-D)}.$

Against the motion vector $t2 = \dfrac{L}{C*(1+D)}.$

Difference of interaction time between the particle and the field

$$t1 - t2 = t.$$

Thus

$$t = \frac{2LD}{C\left(1-D^2\right)}, \quad a = \frac{L3}{t^2}, \quad a = \frac{C^2\left(1-D^2\right)}{2LD},$$

$$F = m*a, \quad F = \frac{mC^2\left(1-D^2\right)}{2LD},$$

$$A = F* L3, \quad A = m*C\char`^2,$$

$$\text{impulse } p = F*t = m*C.$$

If we take into account that the interactions on quantum level will be taken as a second, td = L/C, where L и C const, and motion V/C = D, then

$$F = \frac{mC^2\left(1-D^2\right)}{2tdD}.$$

It also says that the work of a force on the interval (difference of intervals) in a symmetrical field doesn't s depend on the motion and is always locally constant, and the force, and so the acceleration, decrease with the increase of momentum.

3. About Inertia

Here is a simple example. Draw a line. At the middle of it draw a bullet – this will be a particle. Now on both sides from the bullet put small points – elements of the field. These elements move towards the particle.

Speed C = const.

If the particle is stationary (conditionally), then the number of interactions of the particle with field elements will be equal from both sides for any interval that we are calling time. But in fact our particle is moving, and in our example – moving along the line. It has passed an interval equal to the distance between gravitons, i.e. on the drawing – till the first element. Now then, in dynamics in this case the number of interactions against the motion vector is 2 more than along the vector. And this number 2 is true in any part of the Universe.

Let us determine the interval where the particle moves non-uniformly. The number of differences of the interactions of matter (particles) and field on the interval is $S = \dfrac{2\pi * R}{\alpha}$, n = 2, which means that acceleration (negative) against the speed vector in a symmetrical field at the motion of a particle on the interval create 2 non-symmetrical elements of field. And this means that for n = 2 the interval is equal to $S = \dfrac{2\pi * R}{\alpha}$, R = 0,9...10^-15M.

Symmetrical field is such a field whose opposite frequency characteristics are equal - quantity/second.

I.e. that the particle's motion consists of equal intervals, where the particle moves uniformly, regardless of the momentum. Where α is a constant of a thin structure.

Parameters of the particles at the beginning and the end of the interval are equal.

I.e. the forces composed of the difference of forces of counter elements of the field are compensated by two forces free of difference in the motion interval.

But in the moving connected systems of particles frequency of elements fields (quantity / sekond). It is redistributed so, that the frequency characteristic on and against a vector of movement is equal. On a vector increases on 1/1 - D and decreases, as accepted, on (1 - D) and on the contrary. Therefore work of force $m * C^2$ on an interval is compensated to opposite forces. And movement in system begins with zero.

The number of differential forces in the interval is $\dfrac{1}{D} - 1 = \dfrac{1-D}{D}$. Differential force is

$$F = \frac{mC^2\left(1 - D^2\right)}{2tdD}$$

Work of forces on interval S is equal to force F for the sum of intervals

$$L3 = \frac{2L*D}{1-D^2}$$

The sum of intervals L3 =.1 – D /D

$$F3 * \frac{2L}{(1+D)} = m * C^2$$

Two forces balancing them on the interval S are

$$L2 = L\frac{1}{1+D}, \quad t2 = \frac{L}{C*(1+D)}, \quad a = \frac{L2}{t2^2} = \frac{C^2*(1+D)}{L},$$

$$F2 * \frac{L}{1+D} = m * C^2, \quad F4 = 2*F2, \quad F3 = F4.$$

The sum of intervals = 2

F4 *L4 = F2 * 2*L2

F3 = F4.

Inertial force shows only as the motion changes, i.e. at the change of frequency characterisics of the field. If the particle at the interaction with identity elements moves and interacts proportionally to the field's elements, forming motion on the interval, then during the transition to the next level (macro), the movements on the interval are added, i.e. are their sum. In identical conditions the time intervals of the movement and interaction are equal. The uniform movement is composed of equal intervals of the previous level, where each particle moves non-uniformly.

4. Comparative Characteristics (1)

Aristotle

A moving body stops if the force pushing it ceases to have its effect.

Galileo

The speed once rendered to a moving body will be strictly preserved, if the external causes of acceleration or deceleration are eliminated, - a condition that may exist only on a horizontal plane, because in the case of movement on an inclined plane down there already is a cause for acceleration, and in the case of movement on inclined plane up there is deceleration; from this it follows that the movement on a horizontal plane is eternal.

Newton

Every body continues to be kept in a state of rest or of uniform and rectilinear movement, as long as it will not be compelled by the applied forces to change this state.

Modern

There exist such reference systems, for which a material point in the absence of external impacts (on condition of their mutual compensation) keeps the state of rest or of uniform and rectilinear movement.

Inertia is a property of bodies to keep the state of rest or uniform and rectilinear movement, if the external impacts on it are absent or mutually compensated.

From the present work

Inertia is a property of bodies to keep the state of rest or uniform and rectilinear movement, if the external impacts on it are mutually compensated on an interval.

5. A Little about Mass

The mass of particle depends on local energy characteristic of the field. So

$$F = kg * mo*a,$$

where mo is the mass of particle in ideal. Let us imagine such hypothetical possibility , when the particles are distributed in the space so far, that their interactions are equal to zero; kg is a coefficient depending on energy characteristic of the field; $137^2 - k$ – the number of field's elements by one taxonomic act. (one element of the field in the volume), mo – may be taken relative to the Earth surface. Then kg = 1.

He elements of field have no mass. It does not show, because they do not interact. The mass is shown in a particle that is subject to a force. There is no mass without force. $F = m*a$. If we repeat what we have already said, it follows that the motion is absolute. Interactions and time intervals are relative The laws of Nature in any arbitrary small or large area of the space are identical, and if there the mass is present, then proportionally identical. I.e. they may change in some extend quantitatively, but not changing the essence. All interactions between the particles occur by the field and not by direct contact. Thus each individual particle obeys only to the field. This in the principle of independence. The field affects every particle separately, and as all the particles are locally identical, in gravitational field they acquire identical acceleration, regardless of their number and relative location.

We shall cal connected systems such systems for which redistribution of the field depends on the movement direction? Find the frequency characteristics of the field is perceived in the same way in all directions notwithstanding the movement speed. And if the elements are immobile one to another it is perceived as zero movement, regardless to direction.

This occurs regardless to where are the emitter and the receiver located – in a connected system or not, as motion is absolute. There is no reduction of the rod, there is only reduction or expansion of frequency interval of the signal, proportionate to the motion. There is no time reduction, there is relativity of interactions and reduction or expansion of time intervals of interactions, and so of processes.

6. Levels of Interactions

The Nature of inanimate matter has 4 levels of relations.

The 1st relations level that we know about, is determined by a motion larger that C/C. (Begins only from C/C).

The 2nd level forms a measure of relations – space. The space is the volume created by field form 137^2 depending on the matter distribution in it. If the motion is C/C, there is no interactions.

The 3rd level of relations FIELD (not ether)) forms the SUBSTANCE (particles). Substance, resulting from exchange and feedback field – substance creates new qualities in the field, due to purposeful movement equal to V/C. The interactions appear, because the existence of particles implies "in a certain place and at certain time". Interactions are realized through the field resulting from the movement, where the field with const C serves as an "observer". There appears time, as a process related to movement D – movement. There appears mass which is a measure of exchange between particle and field in the space volume.

4th level is macro level, where the movement is speed V. Mass, energy are a sum of components.

Energy is quantitative measure between particle and field in a space unit of movement, i.e. it is the exchange part between particle and field in the space volume.

The particle has exchange mass strictly controlled by the particle itself. It means that the energy of actual field elements received by the particle is used for the energy of inner motion and for dumping the same amount of field's elements. (the mechanism is shown). Thus the energy conservation law is strictly observed.

But a particle interacts not by direct contact, but only by interaction with field, and the field is formed by other particles. The element of field – graviton (gravity) has got a real area of repulsion, which in the interaction with the particle transforms the energy of element(s) into force, work, impulse...., but only into motion or gravity.

Energy is a quantitative measure force, work, impulse.... – is proportional transformation of energy into movement and gravitation. It means that it manifests itself in the form of force, work, impulse...... . Is a carrier.

The field doesn't have mass. The mass manifests itself directly in the interaction between elements of field and a particle.

Energy is a quantitative measure between particle and field.

In ideal the amount of exchanges in a spatial unit of movement is equal to 137^2. But the particle is surrounded by other particles. And although they all are emitting, they partially overlap the amount of exchanges particle-field. (137^2 − k) units. It depends on the density of substance, depth and other factors. The larger is k, the less is the particle's energy, the less is the frequency characteristics of the particle's field.

The spatial unit in this case serves as the length measure $2\pi R*137$, where R is the particle radius, 137 - an inverse value of Vs is equal to 1/137 only in ideal and depends on the particle energy. Spatial unit of time is $\dfrac{2\pi*R*137}{c}$. Frequency characteristics of particle is $\dfrac{Vs}{2\pi*R}$, where Vs = C* α, α − the constant of a thin structure.

Quantitative characteristics of the field does not depend on the movement (gives as much as had received), but, for example, the initial interval of spatial length of photon is equal to Lo and depends also on energy. The more energy − the less is the interval, the less energy the larger is the interval.

Physical objects are field and particle, base of the matter, which participate in processes and in the space.

Space is the measure that forms the dimensions of length, time etc.

The dimension ties by the constants the processes into proportionality and determines the physical identity of processes.

From this the mind "makes conclusions" in the form of physical laws. And rightly so.

7. About Gravitation

The gravitation of any particles system is determined by the difference of the field's frequency characteristics, plus quantitative difference of the external field and the field of larger body, which give the difference of power characteristics. In any particles system there exists partial overlap of the field, though the particle itself is emitting. The energy component particle-field is reduced. 137^2 − k. The Vs of particle is also reduced, and so is also the frequency characteristics of the field. The larger is k, the less is the frequency. For large bodies k increases towards center. It also depends on the substance density. It is typical also for smaller volumes. An example - the work function of electron. But a particle, regardless of the fact that the field may be non-uniform from different directions, emits symmetrically.

A schematic image of a large body, where

1 – a particle with energy state 137^2 – k1,
2 – a particle with energy state 137^2 – k2,
3 – 137^2 – k36
4 – 137^ 2 – k46 k1 < k2 < k3 < k4.
F1 > F2, F2 = F21, F2 > F3. F3 = F31, F3 > F4. F4 = F41 = F5,
where F – is a force proportionate to the frequency characteristics of particles of different levels in a large body.

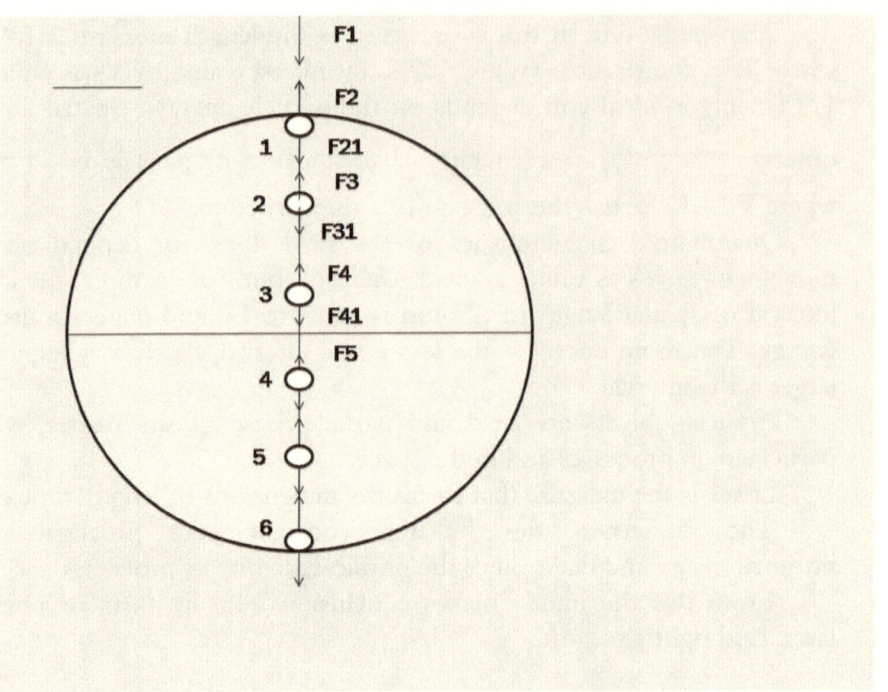

Matter has no infinities, but it has structurally inherent constraints of initial conditions.

C – const*137^2. Vs = 1/137*C. Frequency characteristics of a particle $\frac{C}{2\pi*R*137}$, where 1/137, initial α.

137, 137^2, 1/137 – limiting quantitative characteristics of symmetry.

Each increase of k by a unit gives decrease of α by 3,86082199......* 10^-7 = Ω ,

$$α = 1/137- Ω* k.$$

And so the frequency characteristics, gravitation and initial characteristics of photon depends on local state of α.

$$Vs = α*C, \quad f = \frac{α*C}{2\pi*R}$$

Namely, $to = \dfrac{2\pi*R*137}{C}$, $t = \dfrac{2\pi*R}{\alpha*C}$. I.e., what we mean by time for a given level of matter is in fact a process depending on energy state of the system.

Gravitation of the substance also is determined by the difference of frequency characteristics of fields

$$f1 = \frac{\alpha1*C}{2\pi*R}, f2 = \frac{\alpha2*C}{2\pi*R}, f1 - f2 = \frac{(\alpha1-\alpha2)*C}{2\pi*R},$$

where $(\alpha1 - \alpha2)$ is the difference of frequency characteristics of the substance along the radius from the center of the large body and is equal to

$(\Omega*k1 - \Omega*k2) = \Omega*k,$

$(\alpha1 - \alpha2) = 1/137 - \Omega k1 - 1/137 + \Omega k2 = \Omega(k2 - k 1).$

The force carried by one element of field, not taking into account the difference of these forces and of the motion, is equal to

$F = m*a,$

where L – is a real repulsion area. The force is applied to the particle.

$F = \dfrac{m*C^2}{L}$ - is between the second and the third relations levels.

The fact is that the Nature is divided according to relations levels, and each next level is equal to the sum of previous levels. Thus large numbers of previous level become significantly smaller on the next level. Thus, in the previous formula C/L – frequency from second level to the third, and decreases by frequency characteristics from he third (micro) to the fourth (macro) level. And C^2 – m/sec for 1 - 2 – 3 levels, to C for the third to fourth level.

Frequency characteristics of external field is

$$f1 = \frac{\alpha1*C}{2\pi*R},$$

where $R = 0,9.. *10^{-15}M.$

Frequency characteristics of a particle (sum of particles) at the surface of the body.

$$f2 = \frac{\alpha2*C}{2\pi*R}, f1 = 0,387152*10^{21}*(k2 - k1).$$

The difference of frequency characteristics gives inverse time, during which from the side of external field a force, not balanced from the inner side, is entered. That means that the amount per second is

$$f1 - f2 = \frac{C*137^2(\alpha1-\alpha2)}{2*2\pi*R}, \quad f1 - f2 = \frac{C*137^2*\Omega(k2-k1)}{2*2\pi*R},$$

$f1 - f2 = 0,192313*10^{21}*(k2 - k1).$

The sum of frequency characteristics of the external field with difference of (k2 – k1) with the field of the bigger body, and the difference itself between the external and inner fields, where the difference for one taxonomical act (for one element in the volume) is equal to $137^2/2$ elements of the field. The frequency characteristics here implies the amount per second; k for one particle is a multiple of 1, and the system of particles does not keep this multiplicity.

$$f4 = 0{,}579465*10^{21}*(k2 - k1),$$
$$f3/(k2 - k1) = 0{,}579465*10^{21},$$
$$F = \frac{m*c^2}{L}$$ - the force of identity element of the field, entered

from the side of larger frequency.

$$F = m*a, \quad F = m*\frac{c^2*(k2-k1)}{c*L*f3},$$

a = $5{,}17718*10^3*(k2 - k1)$ - this is proportionate to acceleration of an electron.

a = $2{,}8762*(k2 - k1)$ – acceleration of a proton relative to the mass,

L – real repulsion area of the graviton, equal to 10^{-16} M.

For the Earth the difference of external k1 and k2 (the Earth has k2 approximately equal to 5) at the surface is 3,407… The mass of the fourth level of relations (macro) is equal to the sum of masses of the third level (micro), which gives us the body weight, but regardless of that each particle individually receives the same acceleration.

4. Comparative Characteristics (2)

The special relativity theory of Einstein is based on two postulates, i.e. assertions that are taken to be true within a scientific theory without any proof (in mathematics such assertion are called axioms).

1 postulate of Эйнштейна or relativity principle: all laws of Nature are invariant with regard to all inertial reference systems. All the physical, chemical, biological phenomena proceed in all inertial reference systems identically.

From the present publication

All laws of Nature are invariant with regard to all inertial reference systems. All the physical, chemical, biological phenomena keep their form in all inertial reference systems, and so also their mathematical interpretation, proportionally to the movement and energy within the system, and thus the laws of Nature in moving systems are proportionally identical, because the systems redistribute the frequency characteristics of the field so that it is perceived within the system as the frequency of a body at rest, regardless to this body's movement.

2 postulate or the principle of constancy of the speed of light: speed of light in vacuum in constant and identical with regard to all inertial reference systems. It does not depend on the speed of the light source, nor on the speed of its receiver None of material objects can move with a speed exceeding the speed of light in vacuum. Moreover, none of substance particles, i.e. particle with rest mass different from zero, cannot reach the speed of light in vacuum, such speed may be reached only by field particles, i.e. particles with rest mass equal to zero.

From the present publication

The speed of light in vacuum is constant and identical. C – const. It does not depend on the speed of the light source, nor on the speed of its receiver. Neither does it depend on measure units. None of material objects can move with a speed exceeding the speed of light in vacuum. Moreover, none of substance particles, i.e. particle with rest mass different from zero, cannot reach the speed of light in vacuum, such speed may be reached only by field particles. Speed of field elements (speed of light) does not depend on the movement of the particles system (body), as the dynamics of the field elements is independent. $C/C = 1$.

9. Modern Versions of Michelson's experiment.

In the year 1958 in Columbia University (USA) there had been performed more precise experiment using opposite-directed rays of two masers. The experiment showed invariability of the frequency from the Earth motion up to 10^{-9} %. Even more precise measurements in the year 1974 brought the sensitivity to 0,025 m/sec. The modern versions of Michelson's experiment are using optical cryogen microwave resonators and allow to discover speed of light deviation even if it would be several units by 10^{-16}.

This is one of variants of test in this work (which, as you see, had been performed). It shows that the frequency characteristics (amount/sec) of the field in random tests is unique for the whole manifold. And it is not taken from my head, but from understanding the mechanism of the Nature, which we describe here.

It is one of most fundamental structural features of Nature (together with the limiting parameters of relations structure). Without this phenomenon of Nature, there would be no single reason for preserving the laws of Nature in large or small connected systems, as if they were begun from one initial base. So that the frequency characteristics of uniform movement of particles system is redistributed in such way, that the external field that forms the particles, their motion, "inverse image" made by the particle itself, in any moving system, forms the interactions field-particle in such way, that the symmetry of these interactions shapes a system of relations indistinguishable from a zero motion, and any motion inside the system relative to each other begins from zero, regardless of the general motion.

A simple example. Let us draw a line with two points on it (the distance is not important). These are two moving particles. Vector of their movement is parallel to the line, and the speeds of the particles are equal, which means that relative to both particles their speeds are equal to zero. And if we understand the process of photon forming, then we may practically understand that, regardless of any equal movement of the particles, the frequency characteristics of photon is seen by the particles in the same way, regardless of the motion vector. And not only equal, but also zero movement. (As it actually occurs in the Nature).

There are no postulates and axiomatic statements, but there is an understanding of the structure of relations that are shown in this work. If there is an understanding of the photons structure, frequency characteristics of the field, motion, and their constructivity and self-organization, then one can understand the difference in the comparative characteristics of the assertions based on postulates and axioms – and of the natural mechanism of the Nature.

10. Briefly about Conservation Laws

Impulse field - particle

$$p = m*C.$$

And further the sum by the levels.

The motion on the level field - particle.

$$A = m*C^2.$$

And further the sum by the levels.

But the main thing is not even in this. The Nature may be even wouldn't want to keep the conservation law, but it is organized in such a way, that it cannot not keep it. And the mechanism is rather simple. If the particle receives a certain amount of field elements in one taxonomic act (one element from each of $137^2 - k$ sides, i.e. $137^2 - k$ elements), then exactly the same amount it gives away. The mechanism is shown.

The sum of inner movement (spin), which forms the external relations of a particle with another particle (or a sum of particles), obeys strictly to the external energy laws. It is feedback. $(137^2 - k)$ elements of the field. Increasing k by a unit gives a decrease of the spin's perpendicular component. $(V_s = 1/137C)$ на $3,86082199......* 10^{-7}$ C, which gives a decrease (and for a decrease –gives increase) of the frequency characteristics of the particle and of its photon. (redshift, in particular). Together with the motion and the field's constants (C and L) it gives the physical laws, including the conservation laws.

Here we have given only the main patterns of Nature, and even then by a "not quite scientific" method. There are still many "minor", that need finalization, including technological ones.

11. A Possible Scenario of Birth and Development of the Universe.

The existence and development of Universe could have occurred according to the following scenario. I'll try to tell it based on the considerations given above. The existence of field and gravitons is the original. Also in a field there were such formation as monopoles, that could gather into clusters But they could not form new qualities. Under the influence of gravitation (there had been no other interactions) they compressed and became "thinner". What does it mean "thinner". The monopoles are particles (sacks with quarks), surrounded by the field that also rotates around quark kern. And with sufficient overlapping of the external field, and as a consequence reducing of the covering and reducing of the external "pressure" there is an explosion of the monopoles (most probable in the center) with the quarks flying apart, But the quarks very soon turn into particles. At this time they (particles) actively "take up" the field. The inner field decreases rapidly and as the result of strongly grown gravitation the monopoles are compressed with their destruction and formation of particles. The process is very rapid. When the pressure is balanced, the maim mass remains in the center of the formed galaxy, and the rest scatters and then under the influence of the center's gravitation slows down and moves to orbital motion. But due to fluctuations and uneven destructing process may be compressed and form large monopoles. That are the so-called "Black Holes". Though it is possible that originally the monopoles themselves are the "Black Holes". This process has a chain reaction. As the gravitational background of the formed galaxy has increased, but not more than external one, after some time the neighboring formations are compressed more strongly and there occur the same processes with a certain recession already of the galaxies. We may assume that the relict radiation does not reflect from anything, and this process is going on now with the formation of radiation. The "now" is of course adjusted for the distance. With the formation of particles there appeared new qualities of matter – galaxies, star systems, atoms, planets,.....live matter, man. So the matter has completed its first task. Now I would like to meditate a little about Time. The inanimate matter has got time intervals. Though the matter itself doesn't know anything about it). They are manifested in the motion, in interactions. I.e. in directed processes. But the time intervals of the particles that obey strictly to the field, strictly regulate the constants. And the particles in each case do not have even two degrees of freedom. And this is what causes the strictness of laws in all space.

Time is an universal constant for each identical interaction. It means that the matter cannot perform a part of an action. The Mind with its opportunity has many degrees of freedom. It can perform a part of a "conceived" action. It can control the time intervals in the space. The matter can form many qualities. But it cannot make even such simple things as, say, a wheel (or, rather, four wheels), an ax, a screw, etc. The Mind due to its opportunities can create new qualities. One of such opportunities is to control the time processes. We may make a conclusion – time is a category of Mind. From the previous (page 1) we can conclude, that there is the first circle of feedback: field-matter-field, where the matter is formed by the field, forms new qualities in the field, which create new qualities of the matter. But we may try to conclude also, that there exists one more circle of feedback. What we call Mind – the brain is organized in such a way, that it creates the matter and creates in the field, on the force level, new qualities which   give rise to consciousness, which, in its turn due to the feedback is passed to the live matter, and the matter, now possessing action and time, creates new qualities in the Nature.

 List of literature, and even more - with well-known names of the scientists named in this work, can be considered redundant. Here is a direct solution of the natural problem. This solution has no alternative and this solution yet has not been considered. Also this solution is not alternative to the modern physics, but complements it.

Series: PHYSICS and ASTRONOMY

Khmelnik S.I.

To the Theory of Perpetual Motion Holder

Annotation

The discussed experiment demonstrates the preservation of integrity in a prefabricated construction in the absence of visible fastening forces. It is shown that the experiment can be explained by the fact that a flow of electromagnetic energy appears within the construction. We are considering the conditions in which such flow of electromagnetic energy is maintained for an indefinite length of time.

Contents

1. Introduction

In [1] the following experiment is described – see Fig. 1. Two bars of magneto soft iron with a recess in the middle along all the bar are taken. These bars are put together so that a common canal is formed. A wire is put in it and a current impulse is passed. After this the bars seem to be fastened by a certain force. The force disappears if another impulse is passed though the wire, equal to the previous by its magnitude but opposite by direction. A compulsory condition must be kept – the bars must be tightly adjacent to each other, without any air between them.

The effect cannot be explained by diffusion (as the bars are put together without any pressure and "come off" when the reverse impulse), or by magnetic attraction (as the bars material is magneto soft and does not save the magnetization).

Fig. 1.

The described construction is called in [1] A Perpetual Motion Holder. And I think that such name is very appropriate. Further we shall show that in this construction the thing that is moving – is the flow of electromagnetic energy. Such flow can exist in a stationary system – Feynman in [2] gives an example of energy flow in a system consisting only of an electric charge and a permanent magnet resting side by side.

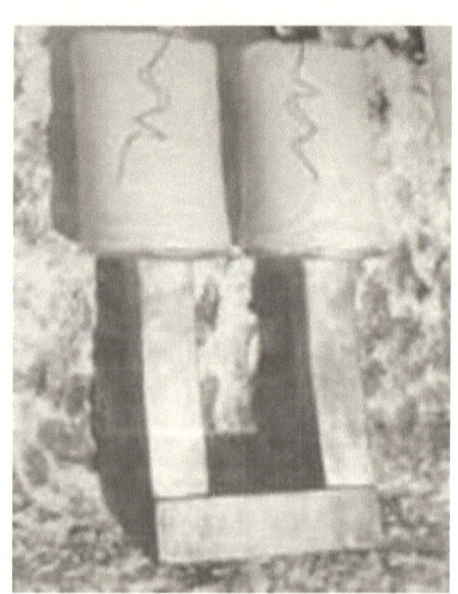

Рис. 2.   Рис. 3.

There exist other experiments demonstrating the same effect [1]. Fig. 2 shows an electromagnet that keeps the attraction force after the current is disconnected. It is assumed that such electromagnets were used by Ed Leedskalnin when building the famous Coral Castle – see Fig. 3 [1].

In all these constructions at the moment of current disconnection the electromagnetic energy has a certain magnitude. This energy may be dissoluted   by radiation and heat losses.   But if these factors are insignificant (at least in the initial period), then the electromagnetic energy should last. In the presence of electromagnetic oscillations a flow of electromagnetic energy must emerge and spread WITHIN the construction. This flow can be interrupted by breaking the construction Then according to the energy conservation law an amount of work equivalent to electromagnetic energy that disappears with the construction destruction, should be expended. It means that the "destroyer" must overcome a certain force. And this is demonstrated in the described experiments.

Below we shall consider the conditions under which the electromagnetic energy can be preserved for an infinite time period.

2. Mathematical Model
2.1. Ferrite Cube

Let us consider a cube, consisting of magneto soft and dielectric material with determined absolute μ and absolute permittivity ε . Let us assume that due to a certain impact in the cube had appeared an electromagnetic wave with the energy W_0 . There is no thermal losses in the , and its radiations (including thermal) are negligible. After some time the wave's parameters will assume stationary values, determined by the values of μ , ε , W_0 and the cube's size. These parameters are the electric field strength and the magnetic field strength as functions of Cartesian coordinates and time, i.e. $E(x, y, z, t)$ and $H(x, y, z, t)$. Naturally, they satisfy the Maxwell equations system of the form:

1.	$\dfrac{\partial H_z}{\partial y} - \dfrac{\partial H_y}{\partial z} - \varepsilon \dfrac{\partial E_x}{\partial t} = 0$
2.	$\dfrac{\partial H_x}{\partial z} - \dfrac{\partial H_z}{\partial x} - \varepsilon \dfrac{\partial E_y}{\partial t} = 0$
3.	$\dfrac{\partial H_y}{\partial x} - \dfrac{\partial H_x}{\partial y} - \varepsilon \dfrac{\partial E_z}{\partial t} = 0$

4.	$\dfrac{\partial E_z}{\partial y} - \dfrac{\partial E_y}{\partial z} + \mu \dfrac{\partial H_x}{\partial t} = 0$	(1)
5.	$\dfrac{\partial E_x}{\partial z} - \dfrac{\partial E_z}{\partial x} + \mu \dfrac{\partial H_y}{\partial t} = 0$	
6.	$\dfrac{\partial E_y}{\partial x} - \dfrac{\partial E_x}{\partial y} + \mu \dfrac{\partial H_z}{\partial t} = 0$	
7.	$\dfrac{\partial E_x}{\partial x} + \dfrac{\partial E_y}{\partial y} + \dfrac{\partial E_z}{\partial z} = 0$	
8.	$\dfrac{\partial H_x}{\partial x} + \dfrac{\partial H_y}{\partial y} + \dfrac{\partial H_z}{\partial z} = 0$	

In order to the cube to have no radiation from the plane xoy, it is necessary that in all points of the plane [3]

$$E_x H_y = 0 \text{ and } E_y H_x = 0. \tag{2}$$

So for a non-radiating cube there must exist a solution of the equation (1), satisfying the said requirement on all the faces of the cube.

Let us consider such solution. We must consider the following functions, (suggested in[4]):

$$E_x(x,y,z,t) = e_x \cos(\alpha x)\sin(\beta y)\sin(\gamma z)\sin(\omega t), \tag{3}$$

$$E_y(x,y,z,t) = e_y \sin(\alpha x)\cos(\beta y)\sin(\gamma z)\sin(\omega t), \tag{4}$$

$$E_z(x,y,z,t) = e_z \sin(\alpha x)\sin(\beta y)\cos(\gamma z)\sin(\omega t), \tag{5}$$

$$H_x(x,y,z,t) = h_x \sin(\alpha x)\cos(\beta y)\cos(\gamma z)\cos(\omega t), \tag{6}$$

$$H_y(x,y,z,t) = h_y \cos(\alpha x)\sin(\beta y)\cos(\gamma z)\cos(\omega t), \tag{7}$$

$$H_z(x,y,z,t) = h_z \cos(\alpha x)\cos(\beta y)\sin(\gamma z)\cos(\omega t), \tag{8}$$

where

$e_x, e_y, e_z, h_x, h_y, h_z$ - the functions amplitudes,

$\alpha, \beta, \lambda, \omega$ - are constants.

Let functions $\sin$ be equal to zero on the cube faces. Let us consider the face xoy for $z=a$ (where a is the length of half-edge of the   cube).   Then $E_x(x,y,z=a)\cdot H_y(x,y,z=a)=0$ and

$E_y(x,y,z=a)\cdot H_x(x,y,z=a)=0$, which follows from (2, 3, 5, 6). It means that this face of the cube does not radiate. Similarly we can show that all faces of the cube do not radiate.

Differentiating and substituting (3-8) into (1); after reducing by common factors, we get:

1.	$h_z\beta - h_y\gamma + e_x\varepsilon\omega = 0$
2.	$h_x\gamma - h_z\alpha + e_y\varepsilon\omega = 0$
3.	$h_y\alpha - h_x\beta + e_z\varepsilon\omega = 0$
4.	$e_z\beta - e_y\gamma - h_x\mu\omega = 0$
5.	$e_x\gamma - e_z\alpha - h_y\mu\omega = 0$
6.	$e_y\alpha - e_x\beta - h_z\mu\omega = 0$
7.	$e_x\alpha + e_y\beta + e_z\gamma = 0$
8.	$h_x\alpha + h_y\beta + h_z\gamma = 0$

$$(9)$$

If $\alpha = \beta = \lambda$, then the equation system (9) takes the form:

1.	$h_z - h_y + e_x\varepsilon\omega/\alpha = 0$
2.	$h_x - h_z + e_y\varepsilon\omega/\alpha = 0$
3.	$h_y - h_x + e_z\varepsilon\omega/\alpha = 0$
4.	$e_z - e_y - h_x\mu\omega/\alpha = 0$
5.	$e_x - e_z - h_y\mu\omega/\alpha = 0$
6.	$e_y - e_x - h_z\mu\omega/\alpha = 0$
7.	$e_x + e_y + e_z = 0$
8.	$h_x + h_y + h_z = 0$

$$(10)$$

In the system (10) equations (10.7, 10.8) follow directly from the previous ones. The first six equations are independent and from equations (10.1-10.6) can be found amplitude functions. We shall seek the solution of system (10.1-10.6) for $h_z = 0$. Then this system will become as follows:

1.	$e_x \varepsilon \omega / \alpha - h_y = 0$
2.	$e_y \varepsilon \omega / \alpha + h_x = 0$
3.	$e_z \varepsilon \omega / \alpha - h_x + h_y = 0$
4.	$-e_y + e_z - h_x \mu \omega / \alpha = 0$
5.	$e_x - e_z - h_y \mu \omega / \alpha = 0$
6.	$-e_x + e_y = 0$

(11)

The solution is:

$$h_y = -h_x,$$ (12)

$$e_x = -\frac{h_x \alpha}{\varepsilon \omega} = 0,$$ (13)

$$e_y = e_x,$$ (14)

$$e_z = -2e_x.$$ (15)

An additional condition for this is the amount of the initial (and stored in the future) energy W_o.

Thus, for a ferrite cube there exists a solution that transforms it into a perpetual motion holder.

2.2. An Iron Cube

Let us consider a cube made from magneto soft iron. In this case it possesses (beside μ and ε) conductivity σ and within it electrical currents of a certain density J may flow. This density should be included into the Maxwell equations – to be more exact, into the first three of (1). Then the Maxwell equations system will become:

1.	$\dfrac{\partial H_z}{\partial y} - \dfrac{\partial H_y}{\partial z} - \varepsilon \dfrac{\partial E_x}{\partial t} + \dfrac{\partial J}{\partial x} = 0$

2.	$\dfrac{\partial H_x}{\partial z} - \dfrac{\partial H_z}{\partial x} - \varepsilon\dfrac{\partial E_y}{\partial t} + \dfrac{\partial J}{\partial y} = 0$	(16)
3.	$\dfrac{\partial H_y}{\partial x} - \dfrac{\partial H_x}{\partial y} - \varepsilon\dfrac{\partial E_z}{\partial t} + \dfrac{\partial J}{\partial z} = 0$	
4-8.	the same as (4-8) from (1)	

The dielectric constant in this case also should have a finite value (as otherwise the electromagnetic wave could not exist). Evidently, the solution satisfying the system (1) will also satisfy the system (16) for $J = 0$. Thus, for a iron cube there exists a solution – an electromagnetic wave without electric current. Of several possible solutions just such solution has to be realized, as for it the electromagnetic wave commits a minimum of work (more precisely – there is no work). So the iron cube turns into a perpetual motion holder.

Conclusion

From the above said it follows that in ferrite and iron cube such electromagnetic wave may be spreading, for which the cube's face do not radiate, and thermal losses are absent (as electric currents are absent even in the iron cube). In these conditions the electromagnetic wave may exist indefinitely –and the cube turns to be a perpetual motion holder. Such cube preserves

- The value of electromagnetic energy,
- The integrity of construction,
- And, possibly, the signal form; in this case we may say that such construction records and stores information.

Apparently, such holder can be made from any magneto soft material and have another, not cubical, form.

References

1. Leedskalnin "Perpetual Motion Holder" (PMH) Bond Effect
   http://peswiki.com/index.php/Directory:Leedskalnin_%22Perpetual_Motion_Holder%22_(PMH)_Bond_Effect
2. R.P. Feynman, R.B. Leighton, M. Sands. The Feynman Lectures on Physics, volume 2, 1964.

3. V.I. Beloded. Electrodynamics. Moscow-Minsk, 2011 (in Russian).

4. S.I. Khmelnik. Variational Principle of Extremum in electromechanical and electrodynamic Systems. Publisher by "MiC", printed in USA, Lulu Inc., ID 1142842, Israel, 2008, ISBN 978-0-557-08231-5.

Series: **PHYSICS and ASTRONOMY**

Khmelnik S. I.

Experimental Clarification of Maxwell-similar Gravitation Equations

Annotation

Maxwell-similar gravitation equations and the experiments of Samokhvalov are considered. It is noted that the observed effects are so significant, that in order to explain them within the said Maxwell-similar gravitation equations these equations should be supplemented by a certain empirical coefficient that may be named gravitational permeability of the medium. Then it is shown that with such supplement the results of experiments are in good agreement with so modified gravitation equations. A crude estimate of this coefficient is given. Some corollaries of these equations are considered, in particular, the gravitational excitation of electric current, the impact of gravito-magnetic induction on the electric current. Some phenomena that can be explained with the aid of these equations are indicated. This paper is a translation of paper [14].

Contents

1. Introduction

There are known Maxwell equations for electromagnetic field of the form (1), proposed by Heaviside [1] (the formulas are given in Appendix 1). Heaviside is also the author the gravitation theory [2], where the gravitation field is described by equations (3) of similar form. Later it was shown [3], that in a weak gravitation field at low speeds from basic equations of general relativity he gravitational analogs of electromagnetic field may be derived, and they have the similar form (3).

Thus, in weak gravitation field of the Earth the Maxwell-similar equations may be used for the description of gravitational interactions. It means that there exist gravitational waves possessing gravito-electric component with intensity E_g and gravito-magnetic component with induction B_g. Mass m, moving in a magnetic field with speed v, is subject to Lorentz gravito-magnetic force (analog of the known Lorentz force) of the form (in the CGS system))

$$F = \varsigma \frac{m}{c} \left[v \times B_g \right] \tag{1}$$

where ς is a coefficient equal to 1 according to Heaviside, and equal to 2 in general relativity theory.

Samokhvalov [4-11] had conceived and carried out a series of unexpected and surprising experiments, which presumably can be explained by interaction of irregular mass currents. Irregular mass currents J_g create variable gravito-electrical intensity E_g and gravito-magnetic induction B_g. At the interaction of this induction with the masses m, moving with speed v there arises gravito-magnetic Lorentz force. It is important to note that the effect are so significant, that in order to explain them within the said Maxwell-similar equations these equations should be supplemented by a certain empirical coefficient ξ. Further it is shown that with such modification the results of experiments are in good agreement with the modified gravitation equations.

Thus, based on the Samokhvalov's experiments the Maxwell-similar equations should be rewritten in the form

$$\text{div}E_g = 4\pi Gm, \tag{2}$$

$$\text{div}B_g = 0, \tag{3}$$

$$\text{rot}E_g = -\frac{\xi}{c}\frac{\partial B_g}{\partial t}, \tag{4}$$

$$\text{rot}B_g = \frac{4\pi G}{c}J_g + \frac{\xi}{c}\frac{\partial E_g}{\partial t}, \tag{5}$$

Where the value of coefficient ξ will be determined below from the said experiments. This coefficient can be called the gravitational permeability of the medium.

2. Certain Analogies and Consequences

Here we shall consider certain analogies between electrodynamics and gravito-electrodynamics, and some consequences of the above examined equations. A qualitative analogy of such kind was indicated by Samkhvalov in [4-11]. One of the consequences has been described in [12].

2.1. The Induction of Circular Mass Current

Magnetic flow Φ, passing through the area S of the coil with the length L, carrying alternated current J, in CGS system

$$\Phi = \frac{4\pi}{c} \cdot \frac{SJ}{L}. \tag{1}$$

The induction average for the area S is

$$B = \frac{4\pi J}{cL}. \tag{2}$$

If the coil is a ring of diameter R, then

$$B = \frac{2J}{cR}. \tag{3}$$

Let us assume now that the ring is carrying alternated mass current J_g.

Then, without considering the technical realization, by analogy with (1.5) we shall get

$$B_g = \frac{2GJ_g}{cR} . \qquad (4)$$

2.2. Gravitational Excitation of Electric Current

From (1.4) follows that gravitational moving force created by gravito-magnetic flow in the circuit of mass current is

$$\varepsilon_g = \frac{\xi}{c} \cdot \frac{d\Phi_g}{dt}, \qquad (5)$$

which differs by coefficient ξ from the similar formula of electrodynamics.

The force of induced electric current in a closed-loop (in the CGS system) is:

$$J = \frac{1}{cR_e} \cdot \frac{d\Phi}{dt}, \qquad (5a)$$

where R_e - the resistance to these electrons motion. This current in the metal is created by free electrons with the charge e_o. By analogy, taking into account (5), we find that variable gravito-magnetic flow Φ_g also creates vortex induced mass current

$$J_g = \frac{\xi}{cR_m} \cdot \frac{d\Phi_g}{dt}, \qquad (6)$$

where R_m is the resistance to mass particles motion This current in the metal is created by free electrons of the mass m_e. Then $R_m = R_e$ - resistance to the electrons motion. In this case mass current J_g corresponds to electric current

$$J_{ge} = J_g \frac{e_o}{m_e} . \qquad (7)$$

It is known that

$$m_e \approx 9.1 \cdot 10^{-34} \text{г}, \quad e_o \approx 1.6 \cdot 10^{-19} \text{Кл},$$

$$\eta = \frac{e_o}{m_e} \approx 1.8 \cdot 10^{14} \frac{\text{Кл}}{\text{г}} . \qquad (8)$$

Thus, the strength of the induced current created by variable gravito-magnetic flow Φ_g is

$$J_{ge} = \frac{\xi\eta}{cR_e} \cdot \frac{d\Phi_g}{dt}. \tag{9}$$

Similarly to (7), the electric current J corresponds to mass current

$$J_{gm} = J\frac{m_e}{e_0}. \tag{9a}$$

Thus, the strength of mass current created by variable magnetic flow Φ is

$$J_{gm} = \frac{1}{cR_e\eta} \cdot \frac{d\Phi}{dt}. \tag{9b}$$

2.3. Rotation of a Porous Ring

Let us consider a ring of average radius R, made of porous metal and charged electrically. Evidently the charges are located on the pores surfaces. Approximally we may assume that the charges distribution density along the ring's circle is described by the function:

$$\rho(\varphi) \approx \rho_0 \cdot (1 + \sin(\lambda\varphi)), \tag{10}$$

where

ρ_0 - is a constant

φ - angular coordinate,

λ - the length of a "wave" depending on the average distance between the pores.

If we cause rotation of the ring with a certain angular speed ω, then the charges distribution density along the circle of the ring becomes a functions of time t

$$\rho(t) \approx \rho_0 \cdot (1 + \sin(\lambda\omega t)), \tag{11}$$

The current flowing in the ring is

$$J(t) = \frac{d\rho(t)}{dt} \approx \rho_0 \cdot \lambda\omega \cdot \cos(\lambda\omega t), \tag{12}$$

where m_0 is a constant. This current creates a magnetic flow that is perpendicular to the ring's plane. The magnetic induction of this flow average by the ring's area is determined in the CGS system by the formula (3). Consequently, the magnetic induction of a rotating charged porous ring, average by the ring's area, is

$$B \approx 2\rho_0 \omega \lambda \cdot \cos(\lambda \omega t)/(cR). \tag{13}$$

By analogy we can state that the rotating porous ring creates mass current

$$J_g(t) = \frac{dm(t)}{dt} \approx m_0 \cdot \lambda \omega \cdot \cos(\lambda \omega t). \tag{14}$$

Then from (4) we find that this current creates variable gravito-magnetic induction

$$B_g \approx 2m_0 G \omega \lambda \cdot \cos(\lambda \omega t)/(cR). \tag{15}$$

2.4. Induction of a Moving Body

It is known that the induction of field created by a charge q, moving with speed $\bar{v}$, in a certain point, is

$$\bar{B} = q(\bar{v} \times \bar{r})/cr^3. \tag{16}$$

The vector $\bar{r}$ is directed from the point, where the moving charge q_1 is located, to the referred point. Similarly, the gravito-magnetic induction of the field created by the mass m, moving with a speed $\bar{v}$, in a certain point, is

$$\bar{B}_g = Gm(\bar{v} \times \bar{r})/cr^3, \tag{17}$$

Because, as shown in the Section 2.2, the electronic current is at the same time also the mass current, the gravito-magnetic induction can create the Lorentz force, affecting the electric current.

2.5. Gravito-magnetic Lorentz Force

In the definition of gravito-magnetic Lorentz force we have used a certain gravito-electric intensity $E'_g = E_g + \frac{\varsigma}{c}\left[v \times B_g\right]$ - see (10) in Appendix 1. By analogy with (4) this expression must be supplemented by a coefficient ξ. The full gravito-magnetic force will get the following definition:

$$F = m\xi\left(E_g + \frac{\varsigma}{c}\left[v \times B_g\right]\right), \tag{18}$$

which also differs from the similar formula in electrodynamics by the coefficient ξ.

2.6. Gravito-magnetic Ampere Force

It is known that a conductor carrying electric current $\overline{J}$ in a magnetic field with induction $\overline{B}$ is affected by Ampere force (per a length unit

$$\overline{F_a} = \frac{1}{c}\left(\overline{J} \times \overline{B}\right) \tag{19}$$

Similarly, a conductor carrying mass current $\overline{J_g}$ in a gravito-magnetic field with induction $\overline{B_g}$ is affected by Ampere force

$$F_{ag} = \frac{\varsigma\xi}{c}\left[\overline{J}_g \times B_g\right] \tag{20}$$

Let us consider the case when mass current is a consequence of electric current, i.e. the particles carrying the charge form the mass current. Then

$$J_g = J\eta_2, \tag{21}$$

$$\eta_2 = m/q, \tag{22}$$

where m, q – mass and charge of the particle. Then a conductor carrying electric current $\overline{J}$ in a gravito-magnetic field with induction $\overline{B_g}$ is affected by Ampere force

$$F_{age} = \frac{\varsigma\xi\eta_2}{c}\left[\overline{J} \times \overline{B_g}\right] \tag{23}$$

For example, if the charged particle is an electron, then

$$m_e \approx 9.1 \cdot 10^{-34}\,\text{г}, \quad e_o \approx 1.6 \cdot 10^{-19}\,\text{Кл},$$

$$\eta_2 = \frac{m_e}{e_o} \approx 0.6 \cdot 10^{-14}\,\frac{\text{g}}{\text{Кл}}. \tag{24}$$

But if the charged particle is an ion with mass $m = h \cdot m_e$, then

$$\eta_2 = \frac{h \cdot m_e}{e_o} \approx 0.6h \cdot 10^{-14}\,\frac{\text{g}}{\text{Кл}}. \tag{25}$$

and for complex molecules $\eta_2 \Rightarrow 1$. So, at the interaction of gravito-magnetic induction with electrical current significant Ampere forces are likely to act.

2.7. Density of Magnetic Wave Energy

It is known that the density of electromagnetic wave energy [13], is

$$W = \frac{B^2}{8\pi} \left[\frac{g}{sm \cdot sec^2} \right] \tag{26}$$

By applying the derivation shown there for the equations (1.2-1.5) of gravito-electromagnetic wave, we find

$$W_g = \frac{\xi^2 B_g^2}{8\pi G}. \tag{27}$$

2.8. Induction of Current-carrying Conductor

It is known that the magnetic induction of infinite conductor carrying electric current is:

$$B = 2J/(cd), \tag{28}$$

where d - is the distance from the conductor to the point of measurement. Similarly, the gravito-magnetic induction of infinite conductor with mass current is

$$B_g = 2GJ_g/(cd). \tag{29}$$

3. Certain Experimental Estimates

The analysis of Samokhvalov's experiments [4-11], performed in Appendix 2, permits to obtain a crude estimate of the coefficient ξ of gravitational permeability. There it was shown that for vacuum

$$\xi \approx 10^{12}. \tag{1}$$

For the air medium this coefficient depends on the pressure. For atmospheric pressure $\xi \Rightarrow 0$, which explains the absence of visible effects of gravitational interaction of moving masses.

There exist several phenomena which can be explained with the aid of the equations (1.2-1.5), considered above, with the existence of coefficient of gravitational permeability ξ – see [12, 15-19].

Appendix 1. The Equations of Electro-magnetism and Gravito-magnetism

Further we shall use the following notations:

- q - electric charge $\left[\sqrt{g \cdot sm}\right]$;

- ρ - electric charge density $\left[\sqrt{g \cdot sm}\Big/sm^3\right]$

- J - electric current density $\left[\dfrac{1}{sm \cdot sec}\sqrt{\dfrac{g}{sm}}\right]$;

- c - speed of light in vacuum $c \approx 3 \cdot 10^{10}\,[sm/sec]$;

- E - electric field intensity
$$\left[\sqrt{g \cdot sm}\Big/sec^2 = 3 \cdot 10^4\,\text{V}/\text{m}\right]$$

- B - magnetic induction $\left[\dfrac{1}{sec}\sqrt{\dfrac{g}{sm}} = Gs\right]$;

- v - speed $[sm/sec]$;

- F - force $\left[dyn = g \cdot sm\Big/sec^2\right]$

- m - mass $[g]$;

- ρ_g - mass density $\left[g\Big/sm^3\right]$

- J_g - mass current density $\left[g\Big/sm^2 sec\right]$

- G - gravitational constant,
$$G \approx 7 \cdot 10^{-8}\left[\dfrac{dyn \cdot sm^2}{g^2} = \dfrac{sm^3}{g \cdot sec^2}\right];$$

- E_g - gravito-electric field intensity $\left[sm\Big/sec^2\right]$

- B_g - gravito-magnetic induction $\left[sm\Big/sec^2\right]$

The Maxwell equations for electromagnetism in CGS system are as follows:

$$\text{div}E = 4\pi\rho, \tag{1}$$

$$\text{div}B = 0, \tag{2}$$

$$(3)\ \text{rot}E = -\dfrac{1}{c}\dfrac{\partial B}{\partial t},$$

$$\text{rot}B = \frac{4\pi}{c}J + \frac{1}{c}\frac{\partial E}{\partial t}. \tag{4}$$

The Lorentz force for the electric charge is

$$F = qE + \frac{q}{c}[v \times B]. \tag{5}$$

The Maxwell equations for gravito-electromagnetism in Gauss CGS system are as follows:

$$\text{div}E_g = 4\pi G\rho_g, \tag{6}$$

$$\text{div}B_g = 0, \tag{7}$$

$$\text{rot}E_g = -\frac{1}{c}\frac{\partial B_g}{\partial t}, \tag{8}$$

$$\text{rot}B_g = \frac{4\pi G}{c}J_g + \frac{1}{c}\frac{\partial E_g}{\partial t}. \tag{9}$$

The Lorentz force for the mass is

$$F = mE_g + \varsigma\frac{m}{c}\left[v \times B_g\right] \tag{10}$$

where ς - is a coefficient equal to 1 by Heaviside and equal to 2 in general relativity theory.

Appendix 2. The Experiments of Samokhvalov
1. Experiment 1

We shall consider the experiment of Samokhvalov described in [4]. Two disks are placed into a vacuum chamber; they are misbalanced (by skewed axes) and are rotating in one direction. Both disks are overheated. Technical parameters of the setup are as follows:

- Material of the disks   aluminum
- Pressure in the chamber $1Pa$
- Density of aluminum $\rho \approx 2.7\text{g/sm}^3$
- Thickness of the disks $h \approx 0.09sm$
- Diameter of the disks $2R = 16.5sm$
- Gap between the disks $d \approx 0.3sm$
- Beating on the sides $0.05sm$
- Number of revolutions $f \approx 50/sec$
- Temperature of overheating (in [4] is written that the temperature rise measured after some minutes was $50K$).

Let us consider the disk's rotation as mass current. We can assume that this current is formed by the mass's motion in the circle of the upper band of the disk of radius $R \approx 7sm$ and the cross-section

$$S \approx 0.3 \cdot 2.5sm^2 \approx 7.5sm^2 .$$ (1)

The speed of this mass is

$$v = 2\pi R \cdot f \approx 2\pi \cdot 7 \cdot 50 \approx 2200sm/\sec .$$ (2)

So, the mass current is

$$J_g = S\rho v \approx 7.5 \cdot 2.7 \cdot 2200 = 4400g/\sec .$$ (3)

This current is variable because the beating of the disks. In accordance with (2.4) this current causes a variable axial induction (along the ox axis of the disk) average on the circle area of radius R,

$$B_g = \frac{2GJ_g}{cR}$$ (4)

or

$$B_g = \frac{2 \cdot 7 \cdot 10^{-8} \cdot 4400}{3 \cdot 10^{10} \cdot 7} \approx 3 \cdot 10^{-15} .$$ (5)

This induction is variable in time because of he disks We shall assume that the circular frequency of this induction is

$$\omega \approx 2\pi f = 314 .$$ (6)

In accordance with (2.9), the strength of vortex electric current created by variable gravito-magnetic flow, is

$$J_{ge} = \frac{\eta\xi}{cR_e} \cdot \frac{d\Phi_g}{dt} .$$ (7)

or

$$J_{ge} = \frac{\eta\xi\omega}{cR_e} \cdot \Phi_g .$$ (8)

In our case

$$\Phi_g = \beta\pi R^2 B_g = \beta\pi R^2 \cdot 3 \cdot 10^{-15} ,$$ (9)

where β – is the coefficient of induction weakening on the level of the driven disk (because of the gap). So,

$$J_{ge} = \frac{\eta\omega\xi}{cR_e} \cdot \beta\pi R^2 B_g$$ (10)

or

$$J_{ge} = \frac{1.8 \cdot 10^{14} \xi \cdot 314}{3 \cdot 10^{10} R_e} \cdot \beta \pi 8.25^2 \cdot 3 \cdot 10^{-15} = \frac{\xi \beta}{R_e} 10^{-6}. \quad \text{(10a)}$$

This electric current raises the disk temperature. In the experiment it was shown that the disk's temperature has increased by $\Delta T \approx 100$ grades Let us consider the equivalent voltage

$$E_e = J_{ge} R_e \quad \text{(11)}$$

And assume that such increase of the disk temperature may be due to the voltage E_e. From (10a, 11) we find

$$E_e = \xi \beta 10^{-6}. \quad \text{(12)}$$

Let us assume that such equivalent voltage is $E_e = 200$. Then we find

$$\xi \beta \approx 2 \cdot 10^8. \quad \text{(13)}$$

Here ξ depends on the pressure, and β depends on the gap. Assuming that $\beta \approx 1/d^2$ and knowing that $d \approx 0.3 sm$, we find $\beta \approx 0.01$. Thus, based on Samokhvalov's experiment we can now assume that for the indicated conditions the gravitational permeability coefficient with the pressure of 0.1 atm is equal to

$$\xi_p(0.1) \approx 2 \cdot 10^{10}. \quad \text{(14)}$$

2. Experiment 2

Let us now consider the experiments of Samokhvalov described in [5]. Two disks are placed into a vacuum chamber, misbalanced by skewed axes. The first of them rotates forcibly, and the second disk begins rotation due to the impact of the first one. The speed f_2 of the second disk's rotation (if the rotation speed of the first one is constant) depends on the gap between the disks d and on the pressure in vacuum chamber p. We may assume that the rotation speed of the driven disk is

$$f_2(p,d) = f_{2p}(p) \cdot f_{2d}(d). \quad \text{(1)}$$

This experiment explores these two dependences.

The dependence of rotation speed on the pressure

$$f_2(p, d = 0.2) = f_{2p}(p) \cdot f_{2d}(0.2) \quad \text{(2)}$$

is given in [5] on Fig 2, from which we find

```
p=[0.1 ,0.3 ,0.5 ,0.7,0.9,1]  (atm),
f=[24, 17, 8, 2, 0.2, ε],
```

where ε is a small value that it is impossible to find from the experiment results. Fig. 1 shows this experimental dependence (by circles) and (by full line) – the approximating function in the form of 5th degree polynomial. In particular, we have

$$f_2(0.1,0.2) = 25, \quad f_2(0,0.2) \approx 37. \tag{2a}$$

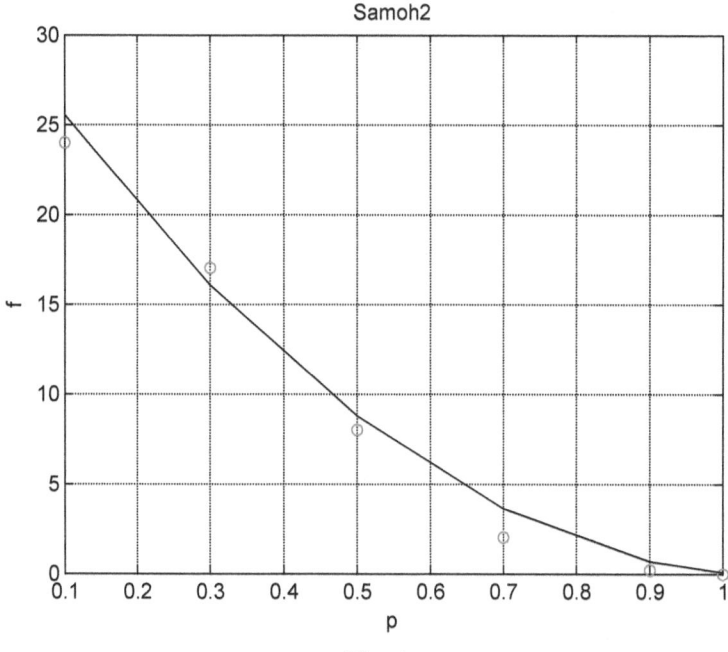

Fig. 1.

The dependence of rotation speed on the distance
is given in [5, Fig. 3], from which we find:

```
d=[0.15, 0.2, 0.25, 0.3]  (sm),
f1=[24, 17, 6, 5] при  p=1atm,
f102=[30, 25, 12, 10] при  p=1.02atm.
```

Fig. 2 shows this experimental dependence (by circles) and the approximating function (by full line) – in the form of $a + b/d^2$, and the function

$$f_{2d}(d) = 1/d^2. \tag{3}$$

To a first approximation further we shall use the function (2). In particular, for $d = 0.3$ (cm) we have $f_{2d}(0.2) \approx 25$.

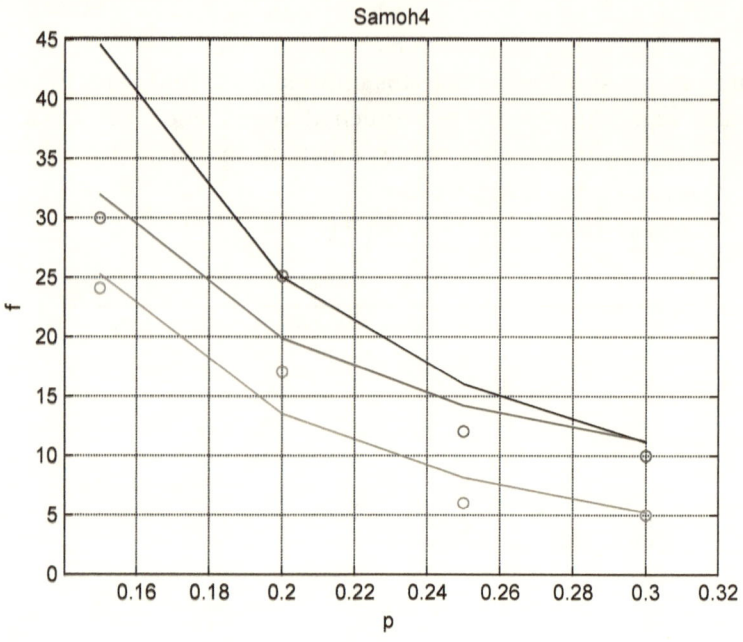

Fig. 2.

Analysis of the functions $f_{2p}(p)$ **and** $f_{2d}(d)$

Taking into account (2, 3), we find:

$$f_{2p}(p) = f_2(p,0.2)/f_{2d}(0.2) = 0.04 f_2(p,0.2). \qquad (4)$$

In particular from (2a) we find:

$$f_{2p}(0.1) = 0.04 f_2(0.1,0.2) = 0.04 \cdot 25 = 1, \qquad (5)$$

$$f_{2p}(0) = 37 \cdot 25 \approx 1000. \qquad (6)$$

From (1, 3, 4) we get:

$$f_2(p,d) = 25 f_2(p, d = 0.2)/d^2. \qquad (7)$$

Below in (3.7) it will be shown that

$$f_{2p}(p) = \vartheta \cdot \xi_p^2(p). \qquad (8)$$

Thus,

$$\xi_p(p) \approx \sqrt{\frac{f_{2p}(p)}{\theta}}, \qquad (9)$$

From (9) it follows that

$$_{(10)}\frac{\xi_p(0)}{\xi_p(p)} \approx \sqrt{\frac{f_{2p}(0)}{f_{2p}(p)}},$$

In experiment 1 it was shown, that

$$\xi_p(0.1) \approx 2 \cdot 10^{10}. \tag{11}$$

Combining (5, 10, 11), we get

$$\xi_p(0) \approx \xi_p(0.1) \sqrt{\frac{f_{2p}(0)}{f_{2p}(0.1)}} \approx 2 \cdot 10^{10} \sqrt{\frac{1000}{1}} \approx 6 \cdot 10^{11}$$

From this we can find a crude estimate of the gravitational permeability of vacuum:

$$\xi \approx 10^{12}. \tag{13}$$

3. The Role of Gravito-magnetic Lorentz Forces

In Samokhvalov's experiments the driving disk drags the driven disk. Now we shall present the explanation of this phenomenon. Samokhvalov notes that first there occurs the vibration of the driving disk, and then begins the rotation of the driven disk – then see Fig. 3.

The disks' vibration is explained in the following way. Above, analyzing the Experiment 1, it was shown that the driving disk is a variable mass current (1.3) with circular frequency (1.6). The "pulsing" mass m_1 creates a variable electro-gravitational intensity

$$E_g = \frac{Gm_1}{d^2}, \tag{0}$$

where d - is the gap between the disks. This intensity is perpendicular to the plane of the disc and on the level of the driven disk affects its mass m_2 by the force (2.18):

$$F_1 = m_2 \xi E_g. \tag{1}$$

Above, when analyzing the experiment 1, we have showed that the masses m_1, m_2 are the mass of a circle of higher band of the disk with radius $R \approx 7 cm$ and cross-section (1.1). This mass is equal to

$$m_1 = m_2 = 2\pi R S \rho. \tag{2}$$

The force F_1 is directed perpendicularly to the disk plane (as and intensity E_g) and varies with the frequency $f \approx 50 / sec$, causing the vibration of the driven disk. Evidently, the speed v_2 of this vibration is proportional to the force F_1, i.e.

$$v_2 = \alpha F_1, \tag{3}$$

where α is a certain constant.

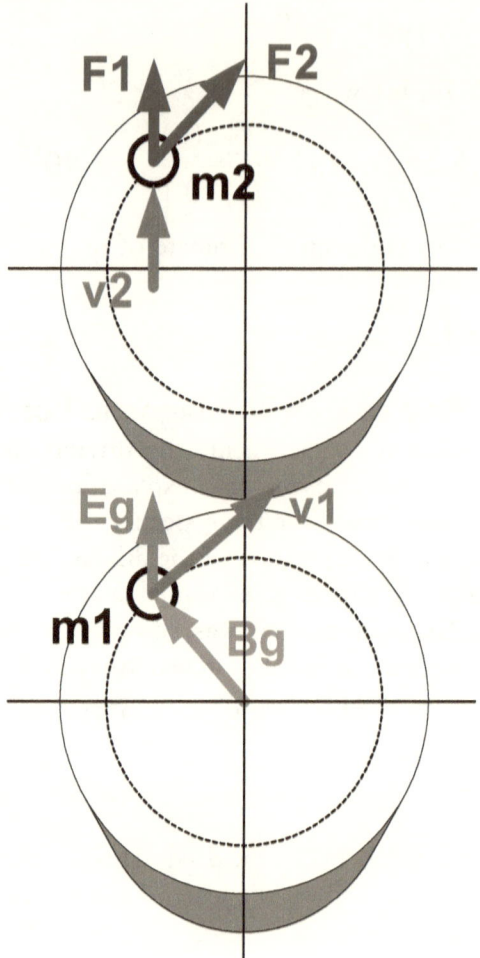

Fig. 3.

This force may explain the "oscillatory" character of the process of repulsion of the screen with the increase of the oscillations amplitude (angle of the frame's deviation) after steadying of the disk rotation speed", which is reflected in the Samokhvalov's experiments described in [8].

Rotating force acting on the driven disk is explained as follows. Gravito-magnetic induction B_g (1.4), created by the driving disk is directed perpendicularly to the mass current of the driving disk, i.e. along the disk's radius and parallel to its plane. This induction acts on the vertically vibrating mass m_2 of the driven disk by gravito-magnetic Lorentz force (2.18):

$$F_2 = m_2 \xi v_2 B_g \frac{\varsigma}{c}. \tag{4}$$

This force is tangential to the circumference of the disc, because perpendicular to the direction of induction (which is directed along the radius of the disk) and the speed (which is perpendicular to the plane of the disk). Due to the fact, that the speed of vibration v_2 and the induction B_g are changing synchronously, the vector of this force doesn't change direction. Apparently, the rotation speed of the driven disk is proportional to the force F_2, i.e. the number of its revolutions is

$$f_2 = \gamma F_2, \tag{5}$$

where γ – a certain constant. Combining (1-5) we get

$$f_2 = \gamma m_2 \xi v_2 B_g \frac{\varsigma}{c} = \gamma m_2 \xi B_g \frac{\varsigma}{c} \alpha F_1 = $$

$$= \gamma m_2 \xi B_g \frac{\varsigma}{c} \alpha m_2 \xi E_g = \alpha \gamma (m_2 \xi)^2 \frac{\varsigma}{c} B_g E_g \tag{6}$$

So, the number of revolutions is

$$f_2 = \vartheta \cdot \xi^2. \tag{7}$$

Which is proportional to ξ^2 with a certain proportionality factor

$$\vartheta = \alpha \gamma m_2^2 \frac{\varsigma}{c} B_g E_g. \tag{8}$$

This ratio is used in the above analysis of the Experiment 2.

References

1. Maxwell's equations. Wikipedia.
2. Oliver Heaviside. A Gravitational and Electromagnetic Analogy. Part I, The Electrician, 31, 281-282 (1893), http://serg.fedosin.ru/Heavisid.htm

3. Gravitomagnetism. Wikipedia.

4. Samokhvalov V.N. Mass-dynamic and Mass-variational interaction of moving masses, "Papers of Independent Authors", publ. «DNA», Israel-Russia, 2009, issue 13, printed in USA, Lulu Inc., ID 7803286, ISBN 978-0-557-18185-8 (in Russian).

5. Samokhvalov V.N. Quadrupole radiation of the rotating masses, "Papers of Independent Authors", publ. «DNA», Israel-Russia, 2010, issue 14, printed in USA, Lulu Inc., ID 8183012, ISBN 978-0-557-28441-2 (in Russian).

6. Samokhvalov V.N. Forceful action of mass-variational radiation on solids. "Papers of Independent Authors", publ. «DNA», Israel-Russia, 2010, issue 15, ISBN 978-0-557-52134-0, printed in USA, Lulu Inc. (in Russian).

7. Samokhvalov V.N. The study of the forceful action and reflection of rotating mass quadrupole radiation from solids. "Papers of Independent Authors", publ. «DNA», Israel-Russia, 2011, issue 18, ISBN 978-1-257-04063-6, printed in USA, Lulu Inc. (in Russian).

8. Samokhvalov V.N. Force effects at mass-dynamic interaction on the average vacuum. "Papers of Independent Authors", publ. «DNA», ISSN 2225-6717, Israel-Russia, 2011, issue 19, ISBN 978-1-105-15373-0, printed in USA, Lulu Inc. (in Russian).

9. Samokhvalov V.N. Dynamic Interaction of Rotating Imbalanced Masses in Vacuum. This issue.

10. Samokhvalov V.N. Research of Force Effects During Mass-dynamic Interaction in Vacuum. This issue.

11. Samokhvalov V.N. Measurement of Force Effects During Mass-dynamic Interaction. This issue.

12. Khmelnik S.I. Detection of gravitational waves. "Papers of Independent Authors", publ. «DNA», ISSN 2225-6717, Israel-Russia, 2012, issue 20, ISBN 978-1-300-07217-1, printed in USA, Lulu Inc., ID 13109103

13. Saveliev I.V. Foundations of Theoretical Physics. Volume 1 - mechanics, electrodynamics. Moskow, "Fizmatgiz", 1991 (in Russian).

14. Khmelnik S.I. Experimental Clarification of Maxwell-similar Gravitation Equations. "Papers of Independent Authors", publ. «DNA», ISSN 2225-6717, Israel-Russia, 2012, issue 21, ISBN 978-1-300-33987-8, printed in USA, Lulu Inc., ID 13325013

15. Khmelnik S.I. The mechanism of origin and method of calculation of turbulent flows, ibid (in Russian).

16. Khmelnik S.I. To the theory of dowsing, ibid (in Russian).

17. Khmelnik S.I. Active field honeycomb, ibid (in Russian).

18. Khmelnik S.I., Khmelnik M.I. Additional forces of interaction of the heavenly bodies, ibid (in Russian).

19. Khmelnik S.I. Sound and gravity, ibid (in Russian).

Samokhvalov V.N.

Dynamic Interaction of Rotating Imbalanced Masses in Vacuum

Annotation

The article presents experimental research results of disk interaction, rotating closely in vacuum, which are not mechanically connected. It was ascertained that at high angular velocities interacting forces between the disks occur, resulting in disk precession and elastic deformation (surface spiral twist) during their conjoint rotation. Meanwhile, a power transfer from the rotating disk to the originally fixed disk occurs, resulting in its rotation, as well as in mutual braking and disk heating during their simultaneous rotation.

Contents

1. Experimental equipment

In our experiments we used a device (Fig. 1) which comprised two direct-current motors D-14FT2s (1 and 2), having electromagnetic brakes, mounted on steel plates 3 and 4 with the thickness of 18 mm.

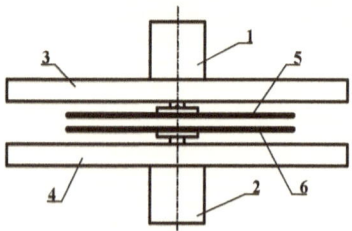

Fig.1. Basic diagram of device for mass-dynamic effect research

The disks (5 and 6) had their diameters of 165 mm and their thickness of 0.9 mm, they were made of aluminum alloy AMg3M (in several cases we used a solid cardboard disk), and they were firmly fixed on the flanges of the motor rotors. The motors were connected to DC power supply B5-48, located outside the chamber in order to maintain a set steady voltage. A separate power supply was used to activate or desactivate the motor electromagnetic brakes.

The distance between the disks was set by parallel movement of electric motor fixing plates on four steel columns, with subsequent rigid fixation. The distance from the disks to the plates was not less than 20 mm. Alongside with that, in our experiments we set both a deliberate distortion of the disk axes relative to their electric motor axes, which created a variable quadrupole moment while the discs were rotating, and the highest possible disk parallelism and dynamic balance were provided. The initial gap between the disks was set from 1 to 6 mm and more. The possibility of mechanical contact during the disk rotation, taking into account their imbalance, was excluded.

The device was installed and rigidly fixed in a vacuum chamber with its inner diameter of 300 mm and its wall thickness of 15 mm (Fig. 2). The air from the chamber was pumped by a vacuum forepump up to residual pressure of about 1 Pa.

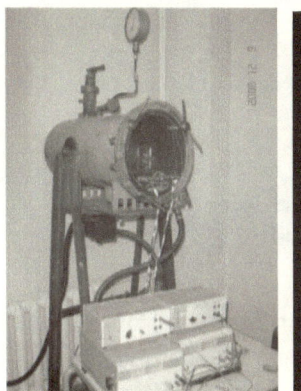

Fig.2. Overall l view of experimental equipment and device in vacuum chamber

2. Experimental research results

In the first series of experiments, the feeding voltage was supplied to both electric motors simultaneously. It is ascertained that if the initial gap between the disks is 1-3 mm (Fig. 3a) and the simultaneous voltage supply is 30 V on both electric motors to rotate them in opposite direction (counter rotation), they are initially accelerated up to a maximum speed of approximately 100-120 1/s (Fig.3b). Then periodical vibration of one of the disks (Fig. 3c), or simultaneous vibration of both disks occurs. (Fig.3d). Disks vibration frequency is approximately 10-20 1/s. During the vibration the disk rotation speed dramatically falls by approximately 2 times (50-60 1/s).

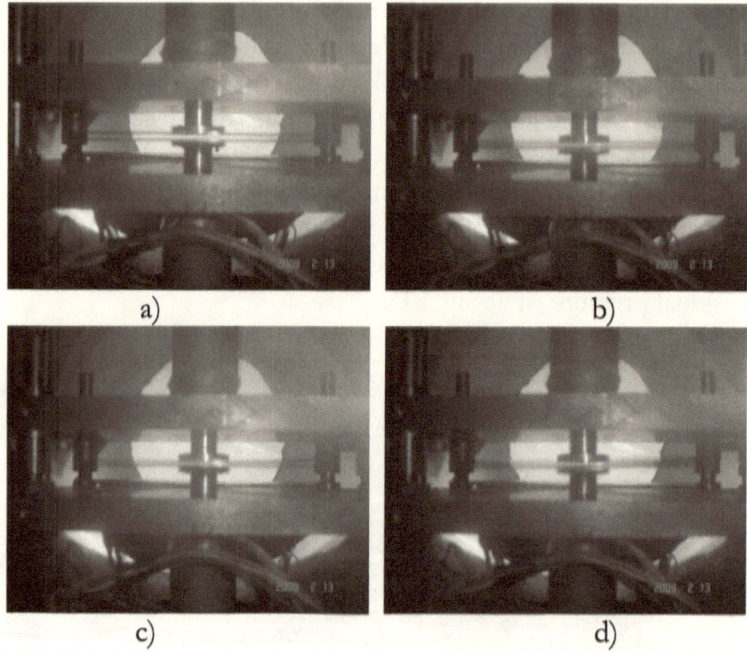

a)   b)

c)   d)

Fig.3. Disk high-amplitude vibration excitation during simultaneous counter rotation in vacuum chamber

During this process, a considerable disk surface bending is observed, i.e. their elastic deformation (Fig. 4). The vibration of one disk is chaotic relatively to the other. The gap between the disk surfaces in different zones is a time-varying parameter.

Fig.4. Disk surface bending during high-amplitude vibration

In this case a mechanical contact between the disks did not occur even if the initial gap between the disks was 1 mm. The disks repelled from each other, which can be seen in the photographs, and each disk was rotating in its own direction. When the vibration stopped the disk rotation speed increased again. The process repeated with certain periodicity.

At some moments the disk chaotic vibrations became relatively stable – the disk spiral twist, rotating with a frequency of 1-3 1/s (Fig. 5).

Fig.5. Disks surface bending and twisting during simultaneous counter rotation in vacuum

In this case, a synchronous distortion of both disk planes occurred. As it can be seen in the photographs and in a slow video of this process, the distorted disk surfaces are almost equidistant. That is, spiral surfaced disks, rotating in opposite directions at a speed of 90-100 1/s, flow around each other without any mechanical contact, a wave of the disk mechanical elastic deformation moves along the surface at the same angular velocity as the disk angular velocity. And the rotation of the spiral twist is in the direction of the disk rotation, which had a higher rotation frequency.

If the disks were deliberately imbalanced (a small disk axes distortion and the rotors axes of electric motors distortion were set), which caused their intense vibration, the above-described effect of the disks interaction (vibration excitation, and then flexural wave occurrence) was observed when the gaps between them were up to 3 mm. In experiments under the same conditions, but in the absence of vacuum (standard pressure in the chamber) these effects did not occur. The strong disk vibration was not excited, and a flexural wave was not observed even when initially set gap between the disks was less than 1 mm.

When one of the electric motors (rotating in vacuum) was switched off, and its disk was stopped, the second electric motor started spinning up to a maximum frequency of approximately 180-200 1/s. When the first electric motor was switched on, the rotating frequency of the second electric motor decreased again. The rotation frequency of both disks was approximately 90-100 1/s. Thereby, during experimental recurrence, it was experimentally ascertained that during conjoint rotation in vacuum a sufficiently strong mutual disk braking was observed.

It was ascertained that in that case during continuous (2-3 minutes), simultaneous, contactless rotation and interaction, the disks are heated to the temperature of 65-70 °C. When the device worked continuously (5-7 minutes), the temperature of the disk heating was up to 80-90 °C. The disk temperature was measured 1-2 minutes after the disks had been stopped and the vacuum chamber had been opened.

While the feeding voltage of 30 V was supplied simultaneously to both electric motors to rotate their disks in the same direction, after complete spinning only a strong vibration of both disks was observed. The distortion of the disk plane bending was not observed.

The electric motor rotation frequency was also significantly less than the maximum frequency. In the process of simultaneous rotation the disks were heated to the temperature of 50-60 °C. When the power supply of one of the electric motors was switched off, the second electric motor spun up to the maximum frequency. When the motor was switched on again, all the effects recurred completely.

If one of the electric motors was switched off and braked, then after the voltage of 30 V was supplied to the second electric motor and after its complete spinning, a slight disk vibration began, and then a slight fixed disk vibration was periodically excited. When the fixed disk vibration was excited, there was a visible decrease in the rotating disk frequency. However, no disk heating even after continuous work was observed.

Thereby, as a result of experimental recurrence, it was ascertained that disc heating occurs only in the case of simultaneous disc rotation in vacuum. The disc heating is the consequence of disc contactless interaction and mutual contactless braking.

In the second series of experiments, the voltage was supplied to only one of the electric motors, while the second one was switched off and braked.

It was experimentally ascertained that if in vacuum a (driven) electric motor was switched off, and braked, when the second (driving) motor was supplied with the voltage of 30V and completely spun, the forced rotation of the first disk with its electric motor rotor began. It was revealed that the effect of excitation of forced rotation and rotation frequency, with other conditions remaining the same, depends on the degree of the disk dynamic balance.

The experiments showed that, with a sufficient high degree of the disks dynamic balance and the absence of disc vibration at a maximum spin, the forced rotation of the driven disk, when the disk gap was more than 2-3 mm, was not excited.

If the gap between disks was 1.0-1.5 mm and the driving disk was completely spun, a slow driven disk turning with a frequency of less than 0.05 1/s was observed. When the driving disk vibration occurred, a driven disk began rotating with a frequency of 5-10 1/s. If the driving disk vibration increased, the driven disc frequency increased up to 20-30 1/s.

At the same time it was ascertained that with a relatively small disk dynamic imbalance the forced disk rotation was excited with the disks gap up to 3 mm. The forced rotation frequency, with other conditions remaining the same, depends on the initial gap between the disks, the less it is, the higher is the rotation speed. When the gap between the disks was more than 4 mm even strong disk vibration did not result in forced rotation excitation of the driven disk.

Thus, the force effect of the driving disk, rotating at a high speed, on mechanically contactless driven disk, resulting in its rotation in vacuum, was experimentally ascertained.

The value of torque, which was developed during this process, was sufficient to rotate the electric motor with the driven disk. The counteraction to this torque, to stop the forced driving disk rotation, required voltage supply to the driven motor, connected with it, which was equal to 0.2-0.8 of the voltage supplied to the electric motor of the driving disk, depending on the gap between the disks and the degree of their imbalance. When the voltage supply of the driving motor was 30 V, to stop the forced rotation of the driven motor with the gap between disks of 1.5 mm, the voltage supply of 12−18 V to the driven disc was required for counter-rotation, and when the gap between the disks was 3 mm, the voltage supply was 5-11 V. When the voltage supply of the driven electric motor was further increased, its disk started to rotate in its own direction (oppositely to the driving disk).

These experiments were carried out without vacuum (under the standard atmospheric pressure in the chamber). While the feeding voltage of the electric motors was the same, the disc rotating speed was lower. In this case, the disk vibration did not occur. The forced rotation of the driven disk was not excited virtually, even when the gap between the disks was less than 1 mm. Under these conditions, only a slow turning of the driven disc with the rotation frequency of less than 0.1-0.3 1/s was observed, i.e. significantly lower than in the case of the driven disk forced rotation in vacuum.

Thereby, as a result of experimental recurrence, it was ascertained that the forced rotation of the initially fixed disk is the consequence of its contactless force interaction with the rotating disk in vacuum. Under these conditions the initial disk precession or vibration preceded its forced rotation. In the absence of vacuum (in the presence of air in the chamber) the forced rotation of the initially fixed disk with a closely located disk, rotating at a high velocity was not virtually excited.

The disks were made of nonmagnetic material and, therefore, other known force interactions were excluded (Barnett effect, etc.).

As the driven disk in our experiments started rotating or its precession was observed, i.e. it received energy from the driving disk, and since the driving disk, with a stable fixed feeding voltage supply of its electric motor, was braked during the driven disk rotation, the driving disk transmitted the part of its rotational energy. The air environment opposed this process.

Conclusion

Proceeding from the analysis of the aforesaid experiment results we can ascertain the following:

1. It was experimentally ascertained that in vacuum the energy was transferred from one (driving) disk, rotating at a high angular velocity to the second initially fixed (driven) disk, which did not have a mechanical

contact with the first disc. At the beginning we observed a precession (or vibration) of the driven disk, then its rotation in the same direction as the driving disk rotated. It was ascertained that the initial disk precession or disk vibration is the necessary condition of the intensification of its forced rotation (when the motor is switched off).

2. It was experimentally ascertained that in vacuum the driving disk rotating at a high velocity has a significant force effect on the closely located driven disk which did not have a mechanical contact with the driving disc. The value of the torque, produced in this process, was sufficiently great not only to rotate the motor with the driven disk, but even to result in the rupture of the disk suspension. When the disk gaps were small, the counteraction to the torque required the voltage supply to the connected electric motor, which was equal to 0.3-0.8 of the voltage supplied to the driving disk electric motor, depending on the gap between the disks and the driving disk imbalance.

3. During a simultaneous high-velocity rotation of the closely located discs, contactless force interaction occurs, resulting in an intense vibration and simultaneous disk distortion, i.e. disk plane bendings, if the disk precession is impossible due to the rigid motor fastening to the electric motor flanges.

4. The force interaction and the disk mutual braking during a long-term contactless rotation in vacuum, result in a significant disc heating (by 50-70 °C). When only one disk rotated no heating was observed.

5. All the effects above mentioned occur when the disks rotate in vacuum. When the disks rotated at a standard atmospheric pressure in the chamber, the disk high-amplitude vibration do not occur, and the plane twist do not occur either. Furthermore, the forced rotation of one disk at a maximum velocity of the other disk rotation is not excited. A slight excitation effect of forced rotation with the frequency less than 0.05-0.1 1/s was observed in the air when the gap between the disks was less than 1 mm.

Series: **PHYSICS and ASTRONOMY**

Samokhvalov V.N.

Research of Force Effects During Mass-dynamic Interaction in Vacuum

Annotation

This article presents the results of the experimental research of rotating mass force effects in vacuum having a variable quadrupole moment on solids. During the research the values of forces, exciting repulsion of solids from rotating mass were measured.

Contents

1. Introduction

In our former experiments it was ascertained that in moderate vacuum (under 0.001 Torr) rotating mass (an aluminum or a cardboard disc) force effects appear, having a variable quadrupole moment on solids which are closely located (e.g. a screen, a disc, a torsion pendulum) both made of nonmagnetic and of conductive materials – mass-dynamic interaction [1-4].

The effect of interaction does not depend on material conductivity of both the disc and the screen and it is followed by the occurrence of electrostatic field or electromagnetic radiation, i.e. it does not have an electromagnetic origin [3, 4]. The necessary condition for a mass-dynamic force interaction occurrence is dynamic (moment) imbalance rotating mass (of the disc), i.e. a variable quadrupole moment.

The main afore experimentally ascertained mass-dynamic interaction force effects in moderate vacuum are as follows:

• Exciting of a flexural wave and a "flow around" rotating in opposite direction dynamically imbalanced discs located with original geometrical clearance of 1,5…3 mm resulting in residual deformation of the fine aluminum disc surface. See:
http://www.youtube.com/watch?v=-O_PnrAa1lM&feature=relmfu;

• Exciting of the disc torsion pendulum rotation with different disc surface orientation relative to the surface of rotating dynamically imbalanced disc (at a distance under 120 mm). See:
http://www.youtube.com/watch?v=r94Lr2CiyBo&feature=relmfu;

• Exciting of forced rotation of a "driven" disc located with geometrical clearance of 1.5…4 mm from the dynamically imbalanced "driving" disc and to stop which it is necessary to supply the voltage of 0.3…0.8 from the voltage feeding of the electric motor rotating the "driving" disc. See:
http://www.youtube.com/watch?v=ochBewD6tVw&feature=relmfu
http://www.youtube.com/watch?v=o9bUi1agnYw&feature=relmfu.

It is experimentally ascertained that with the increase of vacuum depth the value of the force effect (the frequency of forced rotation of the "driven" disc) under other equal conditions grows with the increase of vacuum depth (Fig.1);

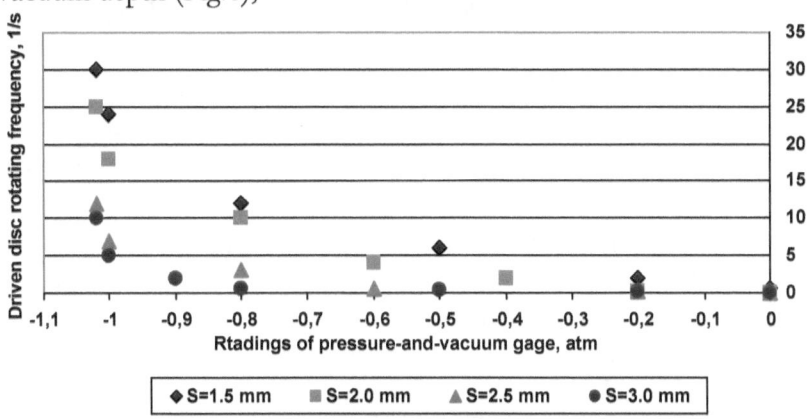

Fig.1. Relationship of frequency of forced rotation of driven disc with clearance (S) between discs and decompression value in vacuum chamber

• Erepulsion ('blowing') of the screen made of fine aluminum foil or stretch film resulting in its irreversible deformation (stretching) and screen material breaking. See Fig. 2, from:
http://www.youtube.com/watch?v=os6naiyT_TU&feature=relmfu

The objective of set forth below series of experiments was the study of the intensity of the force mass dynamic interaction in moderate vacuum (from 0.1 to 0.001 Torr).

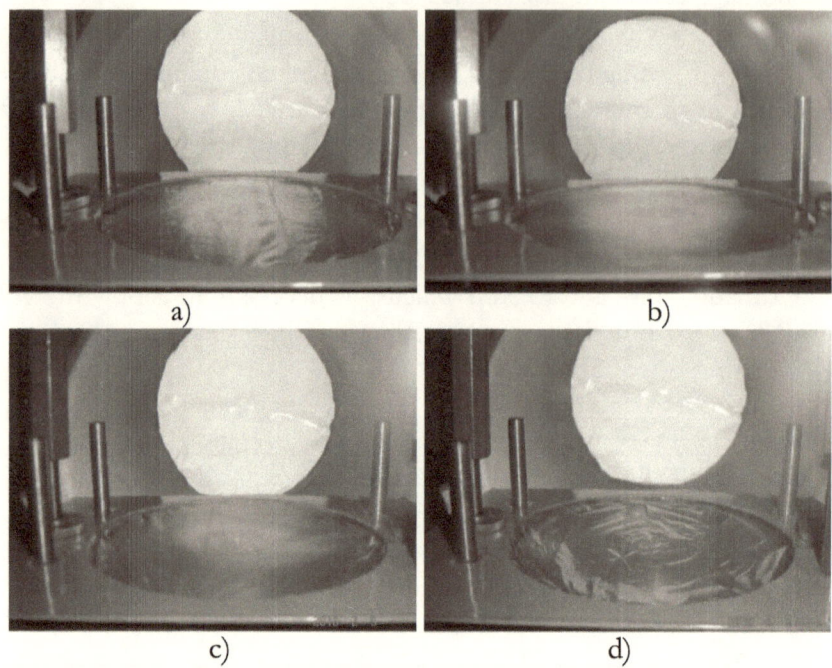

Fig.2. Effects of mass dynamic forces on screen made of stretch film: a) original state, b) screen repulsion from disc, c) vibration waves of stretched screen, d) screen film state after disc was stopped

2. Experimental equipment and tools

The experiments were carried out in the research centre of space energetics of Samara State Aero-Space University (the national research university). The same small vacuum chamber and experimental device were used which the author had used before in the laboratory of Samara State Railway University. But that vacuum chamber was connected to a bigger one at the national research centre (Fig.3.), which has two-stage system of vacuumizing.

Initial air exhaust was carried out by a vacuum forepump HB3-300 up to 0.1 Torr and then some deeper vacuum (up to 0.0008 Torr) was supplied to the chamber by a booster oil-vapor vacuum pump 2HBБM-160. The control and the measurement in the chamber were carried out by a thermocouple vacuum gage BT-2A-П.

Fig.3. Experimental equipment

This gage includes the dynamically imbalanced disc made of aluminum alloy AMr3, with the diameter of 164 mm, the thickness of 0.9 mm, and the mass of 50 gr, being rotated by a direct current electric motor Д-14ФТ2с (U_H=27 V, n=12500 rpm. Electric motor was connected to the source of direct current supply located outside the chamber, which allowed to maintain stable preset voltage. The experimental gage was placed in thrust inside the vacuum chamber. The great thickness of the chamber walls (15 mm) and its great mass together with the rigid placement of the gage almost excluded its vibration while the disc was rotating, which had dynamic (moment) imbalance.

3. Force mass dynamic interaction of rotating disc and rigidly placed screen

The screen was fastened on the rigid and firm console: a steel plate with the cross-section of 5×12 mm (Fig.4.).

The console had an additional support made of bimetallic wire being in contact with the inside surface of the vacuum chamber, it was made to exclude the turn of the console which was clamped by screws. A copper plate with the thickness of 1.3 mm fastened on the cardboard substrate was turned towards the surface of the rotating disc.

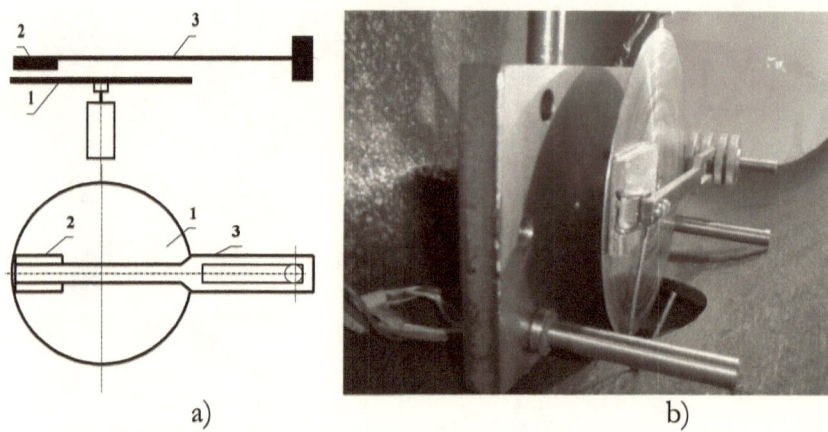

Fig. 4. Basic diagram (a) and overall view (b) of gage with rigidly placed screen: 1 – rotating disc, 2 – screen, 3 – screen fastening console

As the experiments showed while the screen was placed with the clearance of 2 mm from the disc surface, after the beginning of rotation with the supplied voltage U= 30 V of the electric motor, flexural vibration with the amplitude up to 1 mm was excited. Herewith, the vibration intermittently transformed into a flexural wave similar to that which occurred during the interaction of two discs having opposite rotational direction [1]. The amplitude of the flexural wave was about 1 mm under the frequency of its rotation 1...3 1/s (the frequency of the disc rotation was approximately 120...150 1/s). The vibration of the screen was not observed. The mechanical contact of the disc with the screen did not occur. The copper surface of the plate was treated with fine sandpaper to obtain dead surface before each experiment. It was made in order to register traces of mechanical contact of the screen with the rotating disc if they occurred during the experiment.

When fastening the screen with a clearance of about 1,5 mm from the disc surface, after the beginning of the disc rotation with the feeding voltage U=30 V a strong flexural wave was excited. The observed flexural wave frequency on the disc was 1...3 1/s while its amplitude was up to 1.5 mm which resulted in intermittent contact between the disc and the screen. See:

http://www.youtube.com/watch?v=CTF76t3YcA4&feature=relmfu.

When fastening the screen with a clearance of more than 3 mm from the disc surface the described above effect did not occur. After the beginning of the disc rotation even under the high electric motor voltage feeding ($U=$ 35...40 V) and the high disc rotation frequency (up to 180 1/s), a flexural wave on its surface was not excited, i.e. a flexural wave is a consequence of a force interaction of the rotating dynamically imbalanced disc and the screen. The force interaction falls sharply with an increased clearance between the objects owing to a shielding effect of the residual medium in the vacuum chamber like in our former experiments.

Since the screen was almost motionless (in some cases with minor clearances there is a small forced vibration), so it could not generate considerable mass-variational radiation. Therefore the flexural wave excitation was the consequence of the mass-variational interaction (variable mass-variational) field of the rotating dynamically imbalanced disc, with an induced mass-variational field in the screen material.

Since the screen had a small area, so it resulted in local zone effect of the mass-dynamic forces on the screen surface which owing to its relatively low rigidity resulted in flexural wave occurrence on the rotating disc surface.

Without the screen or if it was placed at a distance enough to absorb the energy of the quadrupole (mass-variational) radiation of the residual air medium, the flexural wave on the gyrating disc did not occur.

The value of the force interaction of the rotating dynamically imbalanced disc and the screen resulted in not only excitation of a strong flexural wave. As the measurement of the disc geometry showed after carrying out of 20 experiments (according to above mentioned diagram), initially the flat disc surface transformed into a dome-shaped one (Fig.5.), i.e. a resilient deformation of its material (aluminium alloy AMr3) occurred. The dome height was approximately h=2.4 mm. Since the wall thickness of the disc equaled 0.9 mm, hence the bending arrowhead forming the dome was about 1.5 mm.

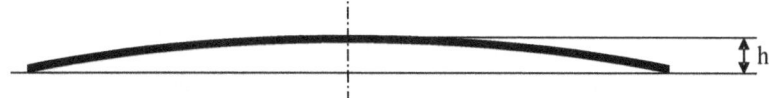

Fig. 5. Disc generating shape after its deformation by quadrupole radiation pressure reflected from screen

The screen copper plate having the thickness of 1.3 mm also obtained a residual deformation. We observed a bending of its corners and there was a bending of its console part located outside the cardboard substrate.

The measurement of the geometry of the discs used before in the experiments with two counterrotating discs showed that the height of their domes was approximately 3 mm, i.e. the bending arrowhead was approximately 2.1 mm. The obtained results are the additional evidence of the considerable value of the mass-dynamic interaction force.

In the experiments with the clearances with the disc (1) movable screens were placed (2), which had the ability of free rotation in the bushes placed on the cardboard substrates (3).

Screen frame "tabs" (2a) were in contact with the cardboard substrates (3) (they rested on it), which excluded a mechanical contact of the screens with the disc (1).

The cardboard substrates (3) (thick cardboard with the thickness of 2.5 mm) also allowed to suppress microoscillations, which could be transferred to the rocker and correspondingly to the screen from the working electric motor and rotating imbalanced disc. Additionally, for account of damping properties of the cardboard a resilient bounce of the screen from the base during their contact in the process of collision was almost excluded.

The cardboard substrates (3) were able to move along spiral columns of the gage, which allowed to place the screens at different distances from the disc with their rigid fixturing. Since the gage itself was placed in thrust in the thick-walled (15 mm) and heavy vacuum chamber it almost excluded the vibration transfer from electric motor through the steel plate of the gage base to its racks and then to the frames (screens).

Initially, two screens were placed above the disc simultaneously (Fig. 6.). The first screen (rectangular frame) was made of bimetallic steel-copper wire with the diameter of 2.4 mm. The second screen (triangular frame) was glued of wooden plates with the width of 10 mm and the thickness of 2 mm.

The wire frame was placed with a clearance of 2 mm with the disc while the wooden frame was placed with a clearance of approximately 3.5 mm.

The clearances are given approximately because the disc had initial axial runout of about 1.5 mm which fell with the increase of the number of the disc rotation revolutions owing to the great centrifugal force effect and its relatively small rigidity (disc thickness was 0.9 mm).

Initially, air pumping was carried out with a vacuum forepump up to residual pressure (0.1 Torr). While feeding voltage (30V) was supplied and the disc was rotated up to 100...120 1/s continuous cyclic vibration of the wire frame was observed at first, as it was placed closer to the disc. The deviation angle of the frame was approximately $\alpha=20°...30°$, the vibration frequency was approximately 4...5 1/s. The vibration of the lighter wooden frame which was placed further from the disc occurred only intermittently. Its deviation angle was up to $\alpha=30°...40°$.

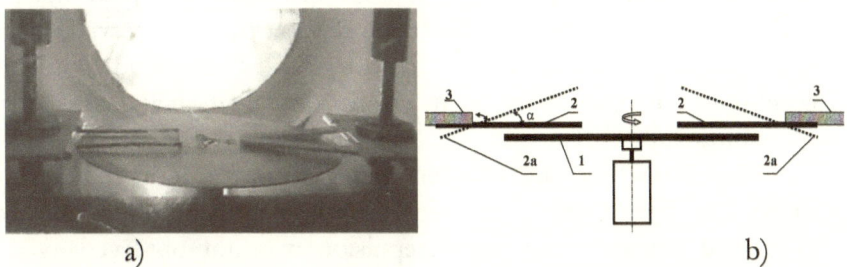

a)   b)

Fig.6. Overall view in vacuum chamber (a) and basic diagram (b) of experimental tool with horizontal screens: 1 – rotating disc, 2 – movable screens (2a – screen frame backing sections "tabs"), 3 – cardboard substrates.

Thereupon without opening the chamber and resetting the gage the air pumping was carried out with an oil-vapor pump up to residual pressure (0.1 Torr). As the experiments showed, the intensity of the mass-variational (quadrupole) radiation force effect of the disc on the screens considerably increased. Continuous vibration was excited in both wire frame and wooden frame. See:
http://www.youtube.com/watch?v=3XF0gCaoqTM&feature=relmfu.
The deviation angle of the wooden frame was up to 70°...80°. The deviation angle of the wire frame was approximately 45°. The greater value of the deviation angle of the wire frame was constructively impossible owing to the frame "'tab" contact with the steel plate of the gage.

In the second series of our experiments the wire frame was placed with a clearance of 3 mm to the disc with a constant setting of the gage and constant electric motor feeding voltage (U=30 V), but with 3 different vacuum values in the chamber: 0.1, 0.01 and 0.001 Torr.

As the experiments showed, with the residual pressure in the vacuum chamber P=0.1 torr, the wire frame repulsion did not occur. When P=0.1 Torr the frame repulsion was excited with a small deviation angle $\alpha = 10°...20°$, while when P=0.001 Torr the repulsion intensity attained the greatest value limited by the "tabs" of the frame. The vibration frequency of the wire frame was approximately 6...10 1/s. Thereby, the intensification of mass-dynamic and mass- variational (quadrupole) radiation force effects with the increase of the vacuum depth in the examined range (from 0.1 up to 0.001 Torr) were ascertained.

In the third series of our experiments both vacuum depth and the distance from the disc varied. The experiment carried out by us showed that when the vacuum depth was increased from 0.1 up to 0.001 Torr, the distance, at which the wire frame repulsion was observed, increased approximately twice from 1.5...2 up to 3.5...4 mm, under other equal conditions (i.e. constant frequency of the disc gyration and constant value of its moment imbalance).

When both wire and wooden frames were placed at the distance of more than 5 mm from the disc and when its gyration frequency was 100...120 1/s the wire frame repulsion was not observed even with P=0.001 Torr.

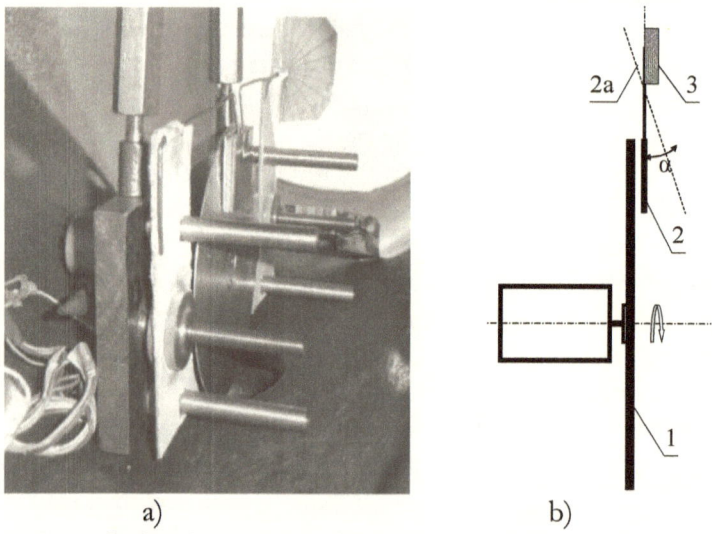

a)   b)

Fig.7. Overall view in vacuum chamber (a) and basic diagram (b) of experimental tool with vertical screen: 1 – rotating disc, 2 – movable screens (2a – screen frame backing sections "tabs"), 3 – cardboard substrates.

The objective of the next series of our experiments was to reveal the impact of the gyrating disc axis spatial orientation and the screen plane location on the process of their non-contact force interaction in moderate vacuum – the vertical screen (Fig.7.). The disc gyration axis was horizontal while the screen was suspended vertically on the wire frame at the distance of 1.5 ...2 mm from the screen plane.

The screen dimensions in the diagram are 50×40 mm, the mass is 16.5 gr. The screen (2) is removable – it is moved on and removed from wire frame work of the rocker (3), which allowed to change the screen plate material turned towards the disc. From the opposite sides of the screen on its base the plates from different materials were fixed. The first plate was made of copper with the thickness of 0.3 mm while the second one was made of aluminum with the thickness of 1.3 mm.

As our experiments showed, when the feeding voltage (U=25 V) was supplied to the electric motor and the disc began to rotate a cyclic deviation of the screen from the disc was observed. The rotary angle of the wire frame was approximately $\alpha=30°...45°$. When the increased feeding voltage (up to U=35 V) was supplied to the electric motor the repulsion angle of the screen increased up to $\alpha=60°...75°$. See: http://www.youtube.com/watch?v=exraNXii2-I&feature=relmfu.

Conclusion

Thereby, it was experimentally ascertained that the rotating disc force effect on the screen occurred independently of the rotating disc axis spatial orientation and the screen plane, i.e. it resulted from the mass-dynamic force effect from the rotating dynamically imbalanced disc.

References

1. Samokhvalov V.N. Dynamic Interaction of Unbalanced Masses in Vacuum. Galilean Electrodynamics, vol. 22, Special Issues 1, 2011, p. 3-6.
2. Samokhvalov V.N. Interaction of Rotating Masses in Vacuum and Their Effect on Solids / Fundamental Problems of Natural Science and Technology. Series "Problems of Universe Research", Issue 34 (M-V). – Snt. Petersburg: Typ. Snt. Petersburg SUCA, 2010. – P. 114-138 (in Russian).
3. Samokhvalov V.N. Mass-dynamic and Mass-variational interaction of moving masses, "Papers of Independent Authors", publ. «DNA», Israel-Russia, 2009, issue 13, printed in USA, Lulu Inc., ID 7803286, ISBN 978-0-557-18185-8 (in Russian).
4. Samokhvalov V.N. Kvadrupolnoye radiation of the rotating masses, "Papers of Independent Authors", publ. «DNA», Israel-Russia, 2010, Issue 14, printed in USA, Lulu Inc., ID 8183012, ISBN 978-0-557-28441-2 (in Russian).

Samokhvalov V.N.

Measurement of Force Effects During Mass-dynamic Interaction

Annotation

This article presents the results of the experimental research of rotating mass force effects in vacuum having a variable quadrupole moment on solids. During the research the values of forces and moments, exciting rotation and repulsion of solids from rotating mass were measured. The mass-dynamic force which was ascertained during our experiments was approximately 2.5...2.7 N while the value of torque was about 1 N·cm. On the grounds of experimentally ascertained force effects, we can assume that a type of interaction under the condition of the presence of a relative mass movement similar to a relative electric charge movement, plays an important role in nature.

Contents

1. Introduction

In our former experiments it was ascertained that in moderate vacuum (under 0.001 Torr) rotating mass (an aluminum or a cardboard disc) force effects appear, having a variable quadrupole moment on solids which are closely located (e.g. a screen, a disc, a torsion pendulum) both made of nonmagnetic and of conductive materials – mass-dynamic interaction [1]. See: http://www.youtube.com/user/Begemotov .

The objective of set forth below series of experiments was the study of the intensity of the force mass dynamic interaction in moderate vacuum (from 0.1 to 0.0008 Torr) and the measurement of the value of occurring forces and torques.

2. Experimental equipment and tools

The experiments were carried out in the research centre of space energetics of Samara State Aero-Space University (the national research university).

Initial air exhaust was carried out by a vacuum forepump HB3-300 up to 0.1 Torr and then some deeper vacuum (up to 0.0008 Torr) was supplied to the chamber by a booster oil-vapor vacuum pump 2HBБM-160. The control and the measurement in the chamber were carried out by a thermocouple vacuum gage BT-2A-П.

This gage includes the dynamically imbalanced disc made of aluminum alloy AMr3, with the diameter of 164 mm, the thickness of 0.9 mm, and the mass of 50 gr, being rotated by a direct current electric motor Д-14ФT2c (U_H=27 V, n=12500 rpm. Electric motor was connected to the source of direct current supply located outside the chamber, which allowed to maintain stable preset voltage. The experimental gage was placed in thrust inside the vacuum chamber. The great thickness of the chamber walls (15 mm) and its great mass together with the rigid placement of the gage almost excluded its vibration while the disc was rotating, which had dynamic (moment) imbalance.

3. Mass-dynamic force effect on the rigid movable screen

In the next series of our experiments we measured the mass dynamic force effect on the rigid movable screen from the rotating dynamically imbalanced disc.

The gage (Fig.1.) was mounted on the frame made of steel angle pieces placed in thrust inside the vacuum chamber with the help of bolts. The steel plate (1) with the thickness of 18 mm is fixed with the bolts on the frame. The direct current electric motor Д-14ФT2c (2) was placed on the plate. The dynamically imbalanced disc (3) made of aluminum alloy AMr3 with the diameter 162 mm, thickness of 0.9 mm, and mass of 50 gr was placed on the electric motor axis. A flat octagonal screen (4) was placed with the adjusted clearance with the disc. The screen was made of thick cardboard with the thickness of 3 mm and it was glued with aluminum foil (0.24 mm).

The screen and the axis (5) were able to move on the axis in the bush. The value of initial spring compression (7) was set with an adjusting screw (8) with a locknut using gauging relation obtained before.

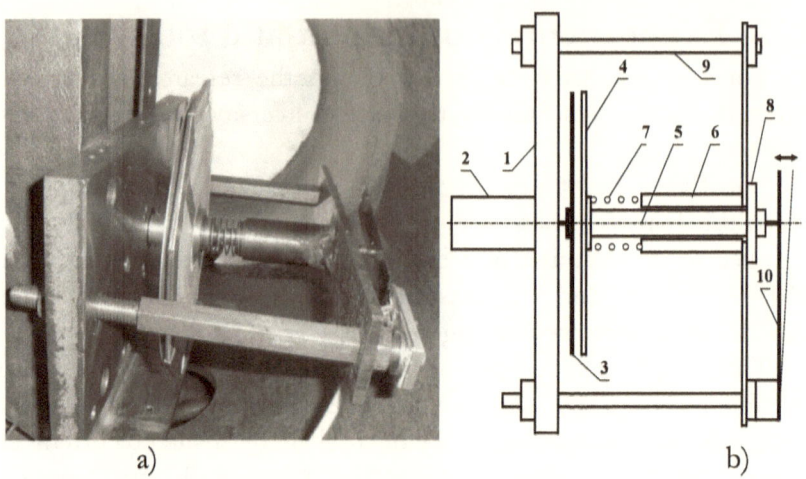

a)   b)

Fig.1. Gage for mass-dynamic value measurement: a – overall view, b) basic diagram

The initial distance from the screen to the disc was set for account of adjusted bar (9). The gage was supplied by a movement strain gauge (10) – resilient console-pinched plate with four strain gauges and adjusted backstop. The indication from the strain gauges when the plate bent (owing to the axial movement) was given to the strain-gauge station with a numeric millivoltmeter. A preliminary gauging of the strain gauge was carried out and we obtained the following:

1) dependence of registered by the strain-gauge station strain on the value of the screen axial movement (the gauging was carried out with the use of the detecting head);

2) dependence of the strain on the force value affecting the screen which is necessary to repel the screen (using a mechanical dynamometer, its spring gauging had been carried out with the help of МИП-1 before).

The force value during the screen axial movement was comprised of the force necessary for the spring compression (7) taking into consideration its preliminary compression (it was set as 2 N), the force necessary for resilient bend of the displacement sensor plate and friction forces between the axis and the bush.

Initially the experiments were carried out with the vacuum depth of 0.1 Torr. The initial clearance between the disc and the screen was set as approximately 3 mm, which securely excluded their mechanical contact. Under the feeding voltage of 30 V and after the disc began to rotate with up to 100...150 1/s the strain gauge indicated the screen repulsion from the disc (displacement from the initial location) to the distance of approximately 1...1.5 mm. As the force of the initial spring compression equaled 2 N, then taking into consideration gauging dependences, the repulsion force – mass dynamic force affecting the screen was in that case approximately 2.2...2.5 N.

Under the feeding voltage of 30 V and after the disc began to rotate with up to 100...150 1/s the disc forced rotation with the frequency of approximately 0.5...1.0 1/s was also excited.

The effect of the screen forced rotation excitation during non-contact mass-dynamic interaction with the rotating dynamically imbalanced disc was similar to that which had been observed during the interaction of the discs [1]. The induced torque in the screen material was sufficient for overcoming the friction force in the axis and the bush and the friction force which was produced by preliminary compressed spring in the rotating and non-rotating parts of the experimental gage being in contact.

During the "pouring" of air into the vacuum chamber to obtain atmospheric pressure without opening the chamber and resetting the gage with the same electric motor feeding voltage the force effects on the screen with the rotating disc did not occur.

Then without resetting the gage the air pumping in the chamber was carried out up to 0.0008 Torr. With feeding voltage of 30 V after the beginning of the disc rotation up to 100...150 1/s the displacement strain gauge indicated the screen repulsion from the disc to the distance of approximately 1.5...2 mm. mas- dynamic force effect on the screen in that case equaled approximately 2.5...2.7 N. The forced rotation frequency increased up to 2...3 1/s. See:
http://www.youtube.com/watch?v=NZaZIKiUEZo

The obtained results confirm that was ascertained before, namely with the increase of the vacuum depth the mass-dynamic interaction value of the disc and the screen increase.

In order to measure the torque rotation, the same experimental gage was used, but the displacement strain gauge (10) was fixed on the gage frame in such a way that its resilient plate was in contact with a dowel (11) which was rigidly fastened with the screen (4) (Fig.2.).

During the screen turn (rotation), the torque excited by the mass-dynamic interaction was defined taking into consideration the bend value of the displacement strain gauge and the moment arm value.

The electric motor feeding voltage and the geometrical clearance between the disc and the screen remained the same (30 V and 3 mm correspondingly). As our experiments showed, under the residual pressure in the chamber of 0.1 Torr the torque value causing the screen forced rotation resulting from the mass-dynamic force effects was approximately 0.5 N·cm. Under the residual pressure in the chamber of 0.001 Torr the torque value was approximately 1 N·cm.

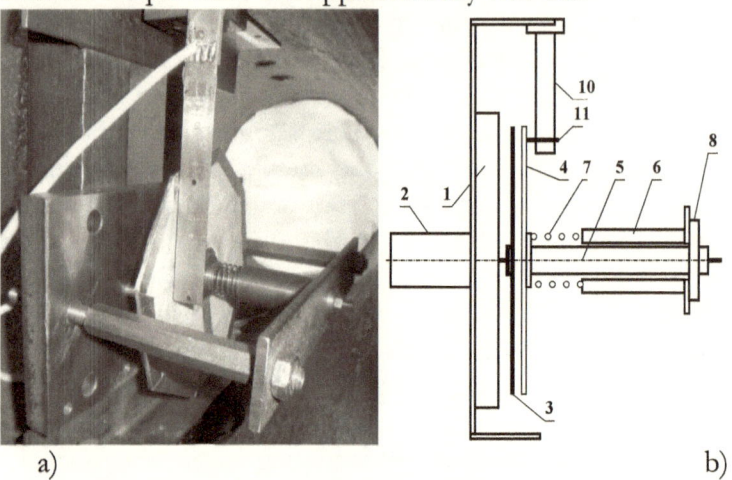

a)   b)

Fig.2. Gage for measurement of mass-dynamic torque value: a – overall view, b – basic diagram

4. The explanation of force mass-dynamic interaction mechanisms

The results obtained experimentally testify that mass-dynamic forces and mass-variational (quadrupole) radiation affects any material objects irrespective of their electric properties (copper and wood). The mass-dynamic force effect has extensional nature similar to the electromagnetic force effect.

The mechanism of the disc mass-dynamic effect (in moderate vacuum) on the screen is as follows. During the rotation of dynamically imbalanced disc each point on its surface and each elemental unit of the disc material rotates on its own circumference (R_i = const), i.e. they do not have their axial movement and, correspondingly, axial acceleration. However, as far as any optional point of the space is concerned (point A,

Fig. 3) which is motionless relative to the disc mass center, where there is test mass m_T, a cyclic approach and drawing of the disc surface (disc mass m_D) determined by the disc rotation frequency ω and the value of its axial runout ΔL take place (Fig.3).

When there is a test mass in point A a relative accelerated movement of the disc mass m_D relative to the test mass m_T, takes place. That is here one should not consider the acceleration applied to the mass (as in Newton's Laws), but they should consider the acceleration of alteration of the distance (location in the space) between masses.

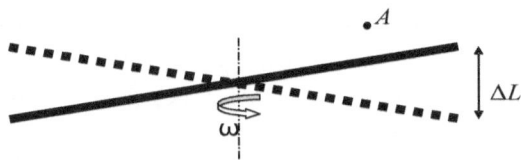

Fig.3. Relative accelerated movement of rotating dynamically imbalanced disc

Thereby, dynamical (moment) imbalance during the disc rotation (variable quadrupole moment) causes a relative accelerated movement of its mass relative to closely located masses (air molecules, screens) which excites mass-variational field, creating a mass-dynamic (spin) polarization of the matter (residual gas medium molecules, screen material molecules).

Mass-dynamic (spin) polarization of the screen material is a vector orientation of the orbital moment of the atomic (molecular) thermal motion quantity of the screen material and also atom spins relative to lines of force of the rotating disc mass-dynamic field (mechanical spin polarization). The necessary conditions for that are mass dynamic field (disc rotation) occurrence and mass-variational (quadrupole) radiation effect (disc dynamic imbalance). The consequence of that is the occurrence of force mass-dynamic interaction of the rotating dynamically imbalanced disc and the screens (discs, torsional pendulums, etc.)

a) **the mechanism of the screen repulsion force occurance**

As one can see in the video record of the process, the screen repulsion begins with a kind of delay after the disc begins to rotate. But then the repulsion continues even with considerable decrease of the disc rotation frequency (after its halt). This can be explained at first by a process delay of the screen material mass dynamic polarization during the disc rotation and then by preservation of the residual polarization of the screen material during a period of time when the disc rotation frequency falls.

Qualitatively, the process of mass-dynamic force effect on the frames made of different materials was the same. But herewith, the repulsion of the heavier wire frame always started earlier than of the wooden frame (during the disc rotation), but it also stopped considerably quicker with the disc rotation frequency decrease. When the frames were placed at the same distance from the disc, the force effect on the wire steel-copper frame was evident to a greater extent (the greater frequency of vibration) than on the wooden frame. It is likely to be determined by different speed and degree of the mass-dynamic polarization of the materials with different density (in this case – cooper and wood).

Vibrational nature of the screen repulsion process under set disc rotation frequency is determined by a greater mass-dynamic force gradient – strong dependence of forces on the distance from the disc and by the decrease of normal constituent of the active mass-dynamic force with the gradient angle alteration of the screen relative to the disc.

Initially, when mass-dynamic forces reached the value exceeding the screen weight, its repulsion from the disc starts, and then owing to the impulse nature of the load application, the screen (frame) passes the part of its trajectory under its own inertia. Thereafter, under the effect of gravitational forces (and also "tab" repulsion from the gage plate – with the wire frame) the screen moves towards the disc, obtains a new impulse, and, thus, the vibration process occurs.

b) the mechanism of the screen rotation excitation

Mass-variational (quadrupole) radiation effect of the rotating dynamically imbalanced disc and its mass-dynamic field result in spin polarization of the screen material. This causes a unidirectional molecule (atom) rotation relative to the lines of force of the mass-dynamic field (their spin orientation).

The molecule (atom) spin orientation excitation – mass-dynamic polarization of the matter result in the fact that (proceeding from retention of the body impulse moment) the whole material object comprised of these atoms, e.g. the driven disc, comes to reverse in direction forced rotation coinciding with the rotary direction of the rotating mass which created a variable mass-dynamic (i.e. mass-variational) field. In this way the driven disc or screen forced rotation is excited with the rotation of the driving dynamically imbalanced disc.

The occurrence of the mass-dynamic field created by a well-balanced rotating disc without mass-variational radiation does not cause spin polarization. This explains that experimentally ascertained force interaction effects occur only with the rotation of the dynamically imbalanced disc (with a variable quadrupole moment) and they disappear with the rotation of the disc not having a dynamic imbalance.

A cyclic nature of the mass-dynamic radiation pressure on each point of the driven disc, set by the driving disc rotation frequency, causes the vibration of the driven disc plane and the precession of its axis to the direction of the driving disc rotation (in the direction of pressure alteration on it, determined by the driving disc rotation). The intensity of mass-variational radiation is relatively small, therefore, as a result of molecule (atom) "pumping" of the disc by the energy of the mass-variational radiation, the process of polarization is prolonged in time. This partly determines the observed in the experiments delay of the screen forced rotation excitation beginning relative to the disc rotation beginning.

5. The analysis of the alternative explanations of observed effects

In the process of discussing the results of the research carried out before, besides effects of electromagnetic nature to explain physical nature of the observed processes in our experiments, two more possible causes were brought forward: 1) the impact of residual air medium; 2) the vibration transferred from the electric motor and the rotating disc.

1) The weakness of the first hypothesis is that the impact of residual air medium resulting in occurrence of the disc force effect on the screen (disc, torsion pendulum) was experimentally proven [2, 3]. It is known that gas-dynamic phenomena with the residual pressure of 0.1 torr almost do not appear, while the effects determined by the gas viscosity sharply decrease with increase of the vacuum depth.

At the same time carried out experiments showed that the force mass-dynamic interaction of the discs and screens (discs, torsion pendulums), on the contrary, increases with the increase of the vacuum depth. This is determined by the decrease of the shielding effect of the residual gas medium.

When the residual pressure is 0.001 Torr (as it was in our experiments), Loschmidt number is sufficiently great (approximately $2 \cdot 10^{14}$ molecules in 1 cm^3), and gas molecules are the same material objects as the molecules of the matter comprising the screen (the second disc or the torsion pendulum). While interacting with variable mass-dynamic field created by the rotating dynamically imbalanced disc, gas molecules orientate their spins (moments of thermal motion quantity, nuclear spins) are oppositely directed to outer field. This creates a shielding effect preventing mass-dynamic field propagation and decreasing its effect on the screen. With vacuum depth increase this shielding effect decreases which was registered in our experiments [3]. Naturally, the shielding effect increases with the increase of the distance from the rotating disc to the screen (disc, torsion pendulum, etc.). This results in value decrease of the force interaction on the screen with the increase of the distance from the rotating dynamically imbalanced disc, which was also registered in our experiments.

This effect has an analogy with electromagnetic interaction – eddy currents induced in electroconductive screen by outer variable magnetic (electromagnetic) field prevent the propagation of electromagnetic field into the depth of material (skin layer, the depth of which depends on material electroconductivity and electromagnetic field frequency).

2) The weakness of the second hypothesis is that vibration effect excited by the rotation of dynamically imbalance disc on force interaction occurrence in the experiments is also obvious.

Firstly, a great mass of the thick- walled vacuum chamber and a massive experimental tool rigidly fixed in it (the total mass is more than 50 kg while the rotating disc mass is 50 gr) almost excludes vibration occurrence. Mechanical oscillation (vibration) of the experimental tool itself in our carried out experiments were not observed.

Secondly and the most important, a great dependence of the observed force interaction on the distance between the disc and the screen was experimentally registered.

A slight alteration of the distance between the disc and the screen (by a few millimeters) resulted in sharp fall, and then to a termination of their force interaction (the screen repulsion and its forced rotation). At that this appeared in experiments with different design, geometry, mass and dimensions of the experimental tool. Such a slight alteration of mechanical system parameters – a slight displacement of the small screen mass (or the second disc) can never result in such an alteration of the system vibrational properties to cause such a great (quantitative and qualitative) alteration of the observed processes nature.

Furthermore, one cannot in principle explain the following experimentally ascertained force interaction effects:

- 'blowing" and repulsion of the screen made of foil and film or repulsion of the rigid screen from the disc with a force of approximately 2.5...2.7 N, which was mentioned above;

- flexural wave excitation and each other "flowing" of the discs rotating in opposite direction or flexural wave excitation on the disc with a rigid placement of a small screen, which was mentioned above;

- forced rotation excitation of the second disc for the halt of which it is necessary to supply the second electric motor with "counter" voltage comparable with electric motor feeding voltage of the first disc or excitation of torque, rotating the screen, with the value of approximately 1 N·cm, which was mentioned above.

6. Force mass-dynamic interaction manifistations in nature

In carried out experiments during vacuumizing as a result of gas medium density decrease, its shielding and masking effect falls. As a result, physical processes determined by the force mass-dynamic interaction and mass-variational radiation start to become clearly apparent. In examined processes force mass-dynamic interaction is observed when the distance is measured in millimeters. Herewith, a sharp rise of the force interaction value with the decrease of the distance between interacting objects (disc, screen) is experimentally ascertained.

A similar interaction, obviously, will take place during any reciprocal relative accelerated mass motions. Force mass-dynamic interaction is manifested in the most dramatic way during gas and liquid mediums motion. Owing to a small distance between interacting molecules of these mediums great mass-dynamic forces appear, which generate (if there is energy inflow) vortical processes (tornado, swirl and so forth) or e.g. short-period low tide or high tide phenomena in the Zhiguli man-made lake [4].

Gravitational waves (they represent mass-variational radiation) from space are not registered on the Earth surface as they are screened (dissipated) by the Earth atmosphere similar to that in our experiments when even a residual air medium screened mass-variational (quadrupole) radiation of the dynamically rotating imbalanced masses.

Electromagnetic interaction determined by the electric charge of particles or bodies, in particular cases are considered as: a) electric field, b) magnetic field, c) electromagnetic radiation. Similar to that mass-dynamic interaction determined by the relative location and relative motion in the space of particle masses or body masses in particular cases are considered as: a) gravitational field, b) mass-dynamic field, c) mass-variational (quadrupole) radiation – gravitational waves [5].

Mass- dynamic interaction is manifested as a matter spin polarization and material body (mass) rotation excitation with their relative motion.

The main observable type of material body motion in nature is their rotation round their own axes (planets, stars) or round the central body (planet systems and so forth). One could assume that the cause of rotation of these bodies and systems is force mass-dynamic interaction of individual particles of the matter which was manifested during their formation in the process of their relative motion under the influence of gravitational forces.

Conclusion

Experimentally ascertained mass-dynamic force affecting the screen was approximately 2.5…2.7 N, while the value of mass-dynamic torque was 1 N·cm.

The manifestation of described above sufficiently great effects of force interaction show that there is a type of interaction in nature determined by a mass similar to that which takes place during a relative motion of electric charges.

In statics it is known as gravitational attraction of masses. In dynamics during a relative motion it is mass rotation excitation, repulsion or attraction of rotating masses – depending on relative orientation of their impulse moments (motion quantity moments). During relative accelerated motion of masses this is excitation of mass-variational (quadrupole) radiation.

References

1. Samokhvalov V.N. Mass-dynamic and Mass-variational interaction of moving masses, "Papers of Independent Authors", publ. «DNA», Israel-Russia, 2009, issue 13, printed in USA, Lulu Inc., ID 7803286, ISBN 978-0-557-18185-8 (in Russian).
2. Samokhvalov V.N. Kvadrupolnoye radiation of the rotating masses, "Papers of Independent Authors", publ. «DNA», Israel-Russia, 2010, Issue 14, printed in USA, Lulu Inc., ID 8183012, ISBN 978-0-557-28441-2 (in Russian).
3. Samokhvalov V.N. Power effects at Mass-dynamic interaction on the average vacuum, "Papers of Independent Authors", publ. «DNA», Israel-Russia, 2011, Issue 19, printed in USA, Lulu IncID 11744286, ISBN 978-1-105-15373-0 (in Russian).

4. Samokhvalov V.N. Research of Vortical Gravitational Field Effect on Air and Water Masses Motion. Written Proceedings of International Forum on the Problems of Science, Technology and Education, vol. 3 / ed. V.A. Malinnikova, V.V Vishnevsky. – M.: Academy of Sciences Of Earth, 2008. – P. 40-41 (in Russian).

5. Samokhvalov V.N. Mass-dynamic and Mass-variational Field in Physical Processes / Fundamental Problems of Natural Science and Technology. Written Proceedings of International Scientific Congress – 2008, issue 33, book II (Н—Я). – Snt. Petersburg: Pearl of Neva, 2008. – P. 473-487 (in Russian).

Zhmud' A.A.

The Nucleons Birth – is the Energy source of Stars and the Sun

Annotation

It is shown that the Nucleons Birth is the Energy source of Stars.

Contents

1. Introduction

From the point of view of the Modern Physical Cosmology, the Energy source of Stars and the Sun is the Nuclear Fusion. In this work it is shown that the Nucleons Birth – is the Energy source of Stars and the Sun.

2. The main Energy parameters of the Sun

The Sun is the Star which is steadily radiates the Energy already on an extent of 3-4 billions years according to objective Geological data. And this radiation is not changed up to now.

Today, the overall Energy radiation of the Sun is: $I_\odot \sim 4 \cdot 10^{33}$ erg/second.

If the Sun radiates such quantity of the Energy already 3-4 billion years, or it is $\sim 10^{17}$ seconds, then the Sun already must to lost about 10^{50} erg as an Energy radiation, i.e.: $E_\odot \sim 2 \cdot 10^{17}$ erg/gram. It is known that the Nuclear Fusion average emitted energy at an elementary act is equal $\sim 10^{19}$ erg/gram [1] that approximately in 50 times more that emitted by the Sun. It also gives the grounds to consider that the Energy source of Stars is the Nuclear Fusion [1].

However the elementary analysis shows that energy of the Nuclear Fusion doesn't suffice for so long existence of our Sun.

3. The Nuclear Fusion is the "unsatisfactory candidate" for the Sun power source

Well, first:

1. It is obvious, what not all mass of the Sun is capable to take part in it to "power career", and only that part which is in a kernel. Moreover, for the general physical reasons, it is possible to assume that in stationary thermonuclear processes of any Star, namely such reaction goes on the Sun any more to exchange of several billions years, at the best no more than 1% of all its weight can participate.

2. According to modern theoretical representations, in stationary thermonuclear reaction all types of thermonuclear transformations have to proceed [2], including there have to be reactions with formation of heavy and super-heavy elements: with mass numbers from 60 to 244 a.e.m. and, probably, more. Such reactions go with big absorption of energy – on the average from 150 to 200 MEV on one atom of a heavy element. It is easy to count that one act of formation of a heavy element with a nuclear weight about 180 a.e.m. withdraws from balance of energy of 23 power acts of elementary thermonuclear synthesis. Let's assume that on 100 acts of elementary thermonuclear synthesis of Helium, there is the only 1 act of formation of any heavy element. Such balance of reactions reduces at once the general balance of energy by 23%. Further, believing that probabilities of formation of heavy elements are approximately equal (it is known already more than 60) it is easy to count that by the end of life of the Sun, the quantity of each of known heavy elements in its structure will be not higher than $1{,}7 \cdot 10^{-2}\%$ in relation to Helium. From here it turns out that if the probability of thermonuclear synthesis of a concrete heavy element on the Sun is at the level of 1/6000, everything together they are capable to take away of themselves about 30% of the energy emitted by the Solar kernel.

Secondly:

1. In works [3-5] it is shown that the substance in the Solar kernel can be only in somebody a degenerate condition, so, no stationary thermonuclear reaction of synthesis, i.e. the Nuclear Fusion, there can exist.

2. By the way, including for the reason of "neutron deficiency" [6].

Well, and most important:

1. According to available empirical data it is known that if for a mix of Hydrogen the Lawson's criterion is carried out, all mix enters thermonuclear reaction in one stage, and in the form of explosion [2].

2. Thus it is known, what in the Sun convective gas processes can't develop [7, 8], so, at once there is a question of how in a zone of the Nuclear Fusion, if it exists, the nuclear fuel arrives and how reaction products are brought out of it?

4. Nucleons Birth – is a power source of Stars and the Sun

According to modern level of knowledge, allocation of energy happens at any processes of transformation of a matter in to a condition to smaller energy. Besides, it is known that the binding energy of the any steady system or object that was emitted at their creation is connected with, and just the same quantity of energy is required for their destruction. Proceeding from specified, it is easy to assume that at synthesis of elementary particles, and in particular nucleons, energy, as, for example, in the case of Nuclear Fusion also has to be emitted. From available empirical data it is known that nucleon binding energy is more than on 2 orders exceed the energy which is emitted in elementary acts of the Nuclear Fusion. Well, and proceeding from it, more logical will be to assume that the Energy source of the Sun is not the Nuclear Fusion, but the Nucleons Birth is.

And it is valid, according to fundamental laws of Quantum Mechanics; the substance in the solar kernel has to pass to a certain degenerate power condition [3-6, 9]. Respectively, the model in which the solar kernel represents the potential well filled with energy in the form of degenerate substance will be the most adequate model of the solar kernel. Such substance, according to the same laws of Quantum Mechanics, will to go out from a potential well in the form of various elementary particles, and nucleons including. And it will to go under probabilistic laws. By estimates, the energy reserved in degenerate substance of the solar kernel is: $E_{\odot я} \sim 1 \cdot 10^{52} - 1 \cdot 10^{60}$ эрг. It is on 30-40 orders more, that already radiated the Sun for the previous history. And the most important is that the specified energy can't be emitted in one stage, in the form of explosion, and only gradually, under probabilistic laws of Quantum Mechanics.

5. Conclusion

It is shown that according to modern level of knowledge, Nucleons Birth is the only "worthy candidate" on the Energy source of Stars.

References

[1]. O. Struve, B. Lynds, H. Pillans. Elementary Astronimy, New York - Oxford, 1959.

[2]. V.I.Boyko. Operated nuclear synthesis and problems of inertial thermonuclear synthesis. Sorosovsky educational magazine, №6, стр. 97-104, 1999 г.

[3]. A.A. Zhmud'. DNA, №15, 2010 г., стр. 114-119. http://dna.izdatelstwo.com/

[4]. A.A. Zhmud'. DNA, №18, 2011 г., стр. 141-145. http://dna.izdatelstwo.com/

[5]. A.A. Zhmud'. DNA, №19, 2011 г., стр. 151-153. http://dna.izdatelstwo.com/

[6]. A.A. Zhmud'. DNA, №19, 2011 г., стр. 154-156. http://dna.izdatelstwo.com/

[7]. Sun riddles. www.tesis.lebedev.ru

[8]. Sun. ru.wikipedia.org/wiki/

[9]. K.N.Mukhin, Experimental nuclear physics, v.1-2, Atomizdat, Moscow, 1974.

Authors

Elkin Igor A. *(Russia)*
IElkin@yandex.ru

Katyshev Alexsey, *Israel.*
kantav@zahav.net.il
Beer Sheva

Khmelnik Solomon I., *Israel.*
solik@netvision.net.il
Ph.D. in technical sciences. Author of about 200 patents, inventions, articles, books. Scientific interests – electrical engineering, power engineering, mathematics, computers.

Samokhvalov Vladimir N., *Russia*
samohvalov_vn@mail.ru
Doctor of Sc., Full Professor, Head of the Department "Construction and Road Machinery and Machine Building Technology" of the Samara State University of Transport.

Zhmud' Alexandr A., *(Russia)*
zalex@sibmail.ru